T0076283

ENVISIONING EXOPLANETS

ENVISIONING EXOPLANETS

SEARCHING FOR LIFE IN THE GALAXY

MICHAEL CARROLL

FOREWORD BY ELISA QUINTANA

ILLUSTRATED BY MICHAEL CARROLL
AND MEMBERS OF THE
INTERNATIONAL ASSOCIATION OF ASTRONOMICAL ARTISTS

This 2020 edition published by Smithsonian Books by arrangement with Elephant Book Company Limited, St. Mary's Hall, Rawstorn Road, Colchester, Essex CO3 3JH

Published by Smithsonian Books
Director: Carolyn Gleason
Senior Editor: Jaime Schwender
Assistant Editor: Julie Huggins

Editorial Director: Will Steeds
Project Editor: Chris McNab
Copyeditor: Magda Nakassis
Designer: Nigel Partridge
Index: James Helling
Color Reproduction: Pixel Colour Imaging Limited, London

This book may be purchased for educational, business, or sales promotional use. For information, please write:
Special Markets Department; Smithsonian Books, P.O. Box 37012, MRC 513, Washington, DC 20013

Library of Congress Cataloging-in-Publication Data

Names: Carroll, Michael W., 1955- author.
Title: Envisioning exoplanets : searching for life in the galaxy / Michael Carroll ; foreword by Elisa Quintana, NASA Goddard Space Flight Center.
Description: Washington, DC : Smithsonian Books, [2020] | Includes bibliographical references and index. | Summary: "Envisioning Exoplanets traces the journey of astronomers and researchers on their quest to explore the universe for a planet like Earth"-- Provided by publisher.
Identifiers: LCCN 2020007210 | ISBN 9781588346919 (hardcover)
Subjects: LCSH: Extrasolar planets. | Habitable planets. | Life on other planets.
Classification: LCC QB820 .C367 2020 | DDC 523.2/4--dc23
LC record available at https://lccn.loc.gov/2020007210

Printed and bound in China, not at government expense
24 23 22 21 20 1 2 3 4 5

Front cover: Image courtesy of Ron Miller.

Half-title: *Superterran (or Super Earth) worlds may range from oversized rocky worlds to small Neptune-like globes. The larger ones may have enough gravity to hold planet-sized moons. This view imagines a gaseous superterran looming over the landscape of a moon with a substantial atmosphere.*

Title page: *A Hot Jupiter orbits within the exotic triple star system of HD 188753. The main star is a Sun-like G star. A duo of smaller stars orbit each other and, in turn, circle around the main star at a distance of over 12 AU, taking 27 years to complete their yearly journey. Within this wild conglomeration, a Hot Jupiter was reported to circle the main star at a distance of only 8 million km (4,970,970 miles), making the circuit every 80 hours. Here, planet HD 188753Ab is seen from a moon.*

BIOGRAPHIES

MICHAEL CARROLL

Award-winning writer, lecturer, and artist Michael Carroll has 30 books in print. Carroll is recipient of the AAS Division for Planetary Sciences Jonathan Eberhart Award for best planetary science article of the year. As a science journalist, he has written for such magazines as *Astronomy*, *Popular Science*, *Astronomy Now* (UK), *Asimov's*, *Analog*, *Sky & Telescope*, *Clubhouse*, *Odyssey*, *Sea Frontiers*, and *Artists*. His books include *Living Among Giants: Exploring and Settling the Outer Solar System*, *Drifting on Alien Winds*, *SPACE ART*, *Alien Volcanoes* (with Rosaly Lopes), and others. His third scientific novel, *Lords of the Ice Moons*, was released in 2018. He co-authored and illustrated the book *Visions of the Revelation* with theologian Jay Adams. During the 2016/17 polar season, Carroll and Rosaly Lopes of JPL embarked on a National Science Foundation-sponsored expedition to study the active Antarctic volcano Mount Erebus and its analogs to ice moons of the outer solar system. Their book, *Antarctica: Earth's Own Ice World* was a 2018 title. Carroll is also a founding member of the International Association of Astronomical Artists (IAAA), and the recipient of the IAAA's Lucien Rudaux Memorial Award for lifetime achievement in the astronomical arts. He has lectured and taught short-term classes at numerous museums, NASA and science centers, and churches. Carroll lives with his lovely wife, Caroline, in Parker, Colorado.

ELISA QUINTANA

NASA astrophysicist Elisa Quintana led the team at NASA's Ames Research Center that discovered Kepler-186 f, an Earth-sized exoplanet in the habitable zone of a red dwarf star. (The habitable zone is the area around a star where a planet could have liquid water on its surface.) Quintana now works at NASA's Goddard Space Flight Center as the deputy project scientist for TESS, a planet-hunting NASA space telescope that is searching the galaxy for Earth-sized planets outside our solar system.

CONTENTS

FOREWORD ... 6

INTRODUCTION ... 8

1 EARLY THOUGHTS AND THE FIRST FINDINGS ... 10

2 KEPLER'S FINDINGS: PLANETS, PLANETS EVERYWHERE, BUT IS THERE ANY LIFE? ... 52

3 CIRCLING A RED DWARF ... 98

4 EXOTIC EXOPLANETS AROUND STRANGER STARS ... 138

5 IT'S AN ALIEN LIFE ... 180

BIBLIOGRAPHY ... 218

INDEX ... 219

ACKNOWLEDGMENTS ... 223

PICTURE CREDITS ... 224

FOREWORD

A planet with two suns in the sky. Lava worlds. Dark planets with no sun in sight. We've read about such worlds in sci-fi books or seen them in movies or on TV. Spine-chilling in a story, awesome on TV, but surely not real? So scientists thought until 30 or so years ago. Since then, thanks to space missions and technological advances in astronomy, scientists now know not only that "exoplanets"—planets beyond our solar system—exist, but there are, literally, thousands of them.

We currently know of more than 4,000 planets; over 4,000 more that are strongly suspected. What has been most exciting about this quest is the dazzling diversity of planets discovered, and the never-ending surprises. Had we known that planets like Jupiter could orbit a star in four days, or that planets were more common around tiny stars, we likely would have discovered them much sooner.

In this book, Michael Carroll charts the remarkable journey that scientists have taken to gain this knowledge, and the critical discoveries they have acquired on the way. He takes us from the early days, when a few lone voices persisted in arguing for the existence of exoplanets, to today, when the thirst to discover more about these strange worlds has become one of the driving forces for modern astronomers.

I first met Michael Carroll while working on the Kepler Mission at NASA Ames. We had many discussions about all of the new and strange planetary systems, and how their orbits and environments might alter their appearances, from the perspectives of standing on the planet itself and also looking down from space.

Fortunately, artists like Michael Carroll can interpret the planet and star data and communicate what is physically there—so very far away—by creating art, the realism of landscapes, skies, and to-scale astronomical nearby objects making you feel as if you are actually visiting these worlds. This art can be so realistic that at NASA we have to label them "artist

concepts," because often a lot of people mistakenly believe they are real photographs. The best part about Michael's art is that it is scientifically driven. Michael takes every clue we have about a system, and pieces it together to illustrate the best representation possible. Dozens of these images are here in this book, alongside those by many other internationally renowned space artists, bringing unimaginably distant landscapes into the realm of our understanding.

Michael's book is certainly timely. As we learn more about stars and planets, new scientific fields evolve and grow that help us understand what these worlds might look like. Astrobiology, the study of life in the universe, involves scientists investigating how different planetary atmospheres filter their starlight, and how that affects what the skies might look like. Others study photosynthesis under different starlight and what colors would be absorbed or reflected by plants. Instead of having blue skies and green foliage, what would planets orbiting red dwarfs look like? What about planets with two different suns in the sky? How might living organisms evolve and adapt to these environments? We also know of many strange new worlds that could never support life, but are fascinating to study. Are these systems really strange, or is it the solar system that is strange? These are questions to think about as you read *Envisioning Exoplanets* and take in the wonderful, scientifically accurate art. Through his own in-depth research and by presenting interview material from some of the world's leading exoplanet astronomers, Michael truly opens up new worlds to the reader and presents a fresh chapter in the history of astronomy.

—Elisa Quintana, 2020

INTRODUCTION

They are lurking out in the dark emptiness, far beyond the limits of our solar system, leaving evidence of their nature and traces of their existence behind. Most of these worlds—known as "exoplanets," or planets that circle other stars—are so distant as to be lost in the glow of their suns, hidden from us. They can scarcely be seen, even with the most powerful telescopes. But scientists want to learn as much as possible about them. We all want to know which of them might be inhabited, possibly raucous and encrusted with living things—but these worlds keep their secrets close to the chest.

Like the investigators at a crime scene, forensic experts are on the trail, sifting through the DNA of the universe, searching out planetary fingerprints in the hearts of stars, in the ebb and flow of starlight, drifting through the nebulae. There, they find the building blocks of life, organic material common to all living things on Earth. In their studies, they also find traces of distant planetary systems. Often, the best clue is a mere shadow, a transient dimming of a star's light as a tiny planet passes before it. Scientists call this event a transit, and if enough transits can be charted, a pattern emerges and observers can decipher a planet's orbit, size, and even mass—all bits of information critical in the search for a life-supporting world. Astronomers craft other techniques in their search, teasing out the shift in a star's color or a telltale wobble in its position, mapping the apparent warp of a background "bent" by an unseen planet's gravity. Even now, researchers continue to invent a host of new approaches.

Somewhere, among the solar systems out there—those of the cool red stars, the more sedate Sun-like stars, or the blazing blue supergiant suns—may lie a very special world. It is a place like home, an "Earth 2.0." But many other exotic worlds keep that hypothetical planet company, planets that even the most imaginative science-fiction visionaries and space artists could scarcely imagine only a few decades ago.

Scientists thought other solar systems would be arranged at least somewhat like our own. We have small, rocky terrestrial planets orbiting closely around the Sun. Beyond them circle the giants: two gas titans (Jupiter and Saturn) and two smaller ice giants (Uranus and Neptune). Farther out are the icy dwarf worlds, unknown to astronomers a century ago. But as the data started to trickle in, and then stream Earthward with the advent of space telescopes, researchers began to get the uneasy feeling that the universe was not so tidy. Yes, there were planets akin to Jupiter and Saturn out there, but they followed crazy orbits, with many circling their stars more closely than Mercury does the Sun. And there were even stranger planets, ones that had no counterparts in our experience. Super Earths, or sub-Neptunes, represented a strange hybrid between Earthlike planets and the ice giants. Some turned out to be huge, stony globes, while others are classified as water worlds, although their water is in a form unlike any Earthly sea. And the astronomers at their telescopes and computer monitors wondered: Which one can harbor life? What form will it take, and will it find anything in common with the life we know on a single, blue planet?

The following pages peer over the shoulders of the planet hunters, past and present: astronomers, philosophers, and today's researchers. Chapters will explore how the very idea that exoplanets existed was at first dismissed; they will delve into the early discoveries, like that of Kepler 10, which was named for NASA's Kepler telescope; they will provide some insight into astronomers' techniques, their day-to-day work, and capture something of the joy of their discoveries. Those discoveries find form in the images taken by the best planet-hunting observatories, and by top astronomical artists who translate the science into landscapes all can understand. Following a historical survey—from those first discoveries, the evolving thinking, and the increasingly sophisticated techniques used in the hunt—*Envisioning Exoplanets* moves on to the wondrous menagerie of worlds scientists have discovered, from blazing gas giants to frigid mini-Earths. Among them, we search for Earth 2.0, a cousin world to our own remarkable blue oasis in space.

The count for confirmed exoplanets is approaching the 5,000-planet mark, but that is only the tip of the iceberg. With all of those worlds out there, one might think astronomers would find at least a few others like ours. Chapter 2 unpacks why we aren't finding more Earthlike worlds and carries out a brief survey of initial discoveries. The following chapter hones in on planets that orbit the most common stars, the red dwarfs (or M stars). The book then moves on to consider a host of exoplanets, focusing on the strange and beautiful in our search for life among the stars. Ensuing pages survey Earth candidates and explore where observers might find life on planets both Earthlike and very alien. Finally, Chapter 5 discusses the future possibilities: What are our chances of exploring such worlds, remotely or with human expeditions? Do the exoplanets teach us lessons to help us understand and care for our home world? And finally, is there life among the exoplanets, primitive or intelligent? It's time to embark on a search for life in the galaxy.

—Michael Carroll, 2020

Right: *A full 90 percent of all stars are cool red dwarfs, or M stars. Giant planets are often found in close orbits around these stars. These planets will undoubtedly have many moons circling them, so their skies will be spectacular. Here, we see a typical scenario, the view of a gaseous giant planet seen from an icy moon, with another moon partially eclipsing the red dwarf sun.*

1

EARLY THOUGHTS AND THE FIRST FINDINGS

Humans have always wondered what was around the corner, what lay just over the next hill, what treasures awaited beyond that horizon. Can we find places of sanctuary, of familiarity? Biologists tell us it's a built-in survival strategy: move out, expand the species, or die. Evolutionists suggest that our curiosity is residue from our early transition from sea to shore. Philosophers tell us that exploration is a hardwired feature of what makes us human. Theologians propose that our quest is for the Other, for someone like us, a search to fill a spiritual void. Perhaps it is all of these, and more. With the advent of space observation and exploration, we find ourselves once again wondering: What treasures are out there? Is there a place like home, an Edenic garden to be explored, a cosmic sanctuary in which to find solace? Every month brings more worlds to ponder as our observatories and spacecraft discover new planets circling distant stars. Among these exoplanets we find many bizarre and intriguing locales. Perhaps among them we will find our place of solace, our "Earth 2.0." Or perhaps we will discover life very different from anything on our planet, but life nevertheless.

Left: *Scorching temperatures heat the atmosphere of the Hot Jupiter WASP-121 b into a frenzy. As the atmosphere is blown away by the solar wind of WASP-121, the gravity of the star pulls the planet into a football shape. This imaginary view is from an orbiting moon, which would suffer the same extreme temperatures that its planet does. Hot Jupiters are one of the most common types of exoplanet found.*

THE STARS IN THE SKY

Our Sun is a G-type star. It was born in a cloud of dust and gas about 4.6 billion years ago. It began as the hub of a great disk-shaped cloud, a central bulge that gained more and more material. As the material at the center accumulated, its gravity increased, pulling in dust and rock to itself, becoming a dense globe of material. Most of the matter making up the Sun and its surrounding cloud was hydrogen, and eventually pressures became so great that the very atoms within the central sphere collapsed, and nuclear fusion began. A star was born.

Because of the type of star it is, it's in middle age. Stars like the Sun burn

LIKE SOME RECKLESS HOLLYWOOD PERSONALITIES, LIVING HOT AND FAST MAY LEAD TO SHORT LIFE.

Above: *Chile's Atacama Large Millimeter Array imaged this protoplanetary disk surrounding the infant star HL Tauri. The image shows details in the disk's structure such as cleared regions (dark) and possible locations of planets forming within the cloud.*

Left: *Stars and their planetary systems begin as vast disks of dust and gas. As the central portion of these disks collapses, a sun ignites while smaller planets form around it.*

Above: *The red giant Betelgeuse has nearly exhausted its hydrogen and is now burning helium. When a star reaches this stage, it expands and enters its final stages of life.*

hydrogen, and that hydrogen becomes helium. As stars continue through this process, they are called "main sequence" stars, which are fundamentally stable—our own star has been steadily burning hydrogen for over 4 billion years and will continue to do so reliably for another 8 billion years. Planetary systems surrounding any main sequence stars are thought to be the best places for life to arise.

A star's lifetime is determined by its size. The bigger the star, the hotter it is and the faster it will transform its hydrogen into helium, finally burning out in a whimper or an explosion. Like some reckless Hollywood personalities, living hot and fast may lead to short life. Blue supergiants have life spans of only millions of years. On the other end of the scale, the M stars or red dwarfs are small and cool, burning their fuel slowly. Many

have been around since the beginning of the universe, and will be the most ancient surviving stars to the end.

As the Sun and other stars exhaust their hydrogen, helium takes over. (The largest stars also burn carbon.) The star begins to expand, becoming a red giant or a supergiant, swelling up to many times its original size. When our Sun becomes a red giant, it will likely consume Mercury, Venus, Earth, and perhaps Mars. One of the most famous red giants is Betelgeuse in the constellation Orion. If placed at the center of our solar system, this behemoth would extend into the asteroid belt. Stars 10 times more massive than the Sun—like Betelgeuse—will eventually drive their outer layers away as the core collapses into a dense, tiny star like a white dwarf or, in the case of larger ones, a neutron star or black hole. The explosion from the collapsing of the core is what causes a supernova. In some cases, the leftover remnant of a supernova may leave behind a dense star corpse like a white dwarf or even a pulsar.

The star types across our galaxy are varied and afford differing probabilities of life-friendly conditions. The most common stars are the red dwarfs, Jupiter-size cool stars making up nearly 80 percent of all suns. These are followed by K-type stars, sometimes called orange dwarfs, which make up one-tenth of the stars in the sky and are just a bit smaller than the Sun. Our Sun joins the other G stars, representing about 7 percent. The

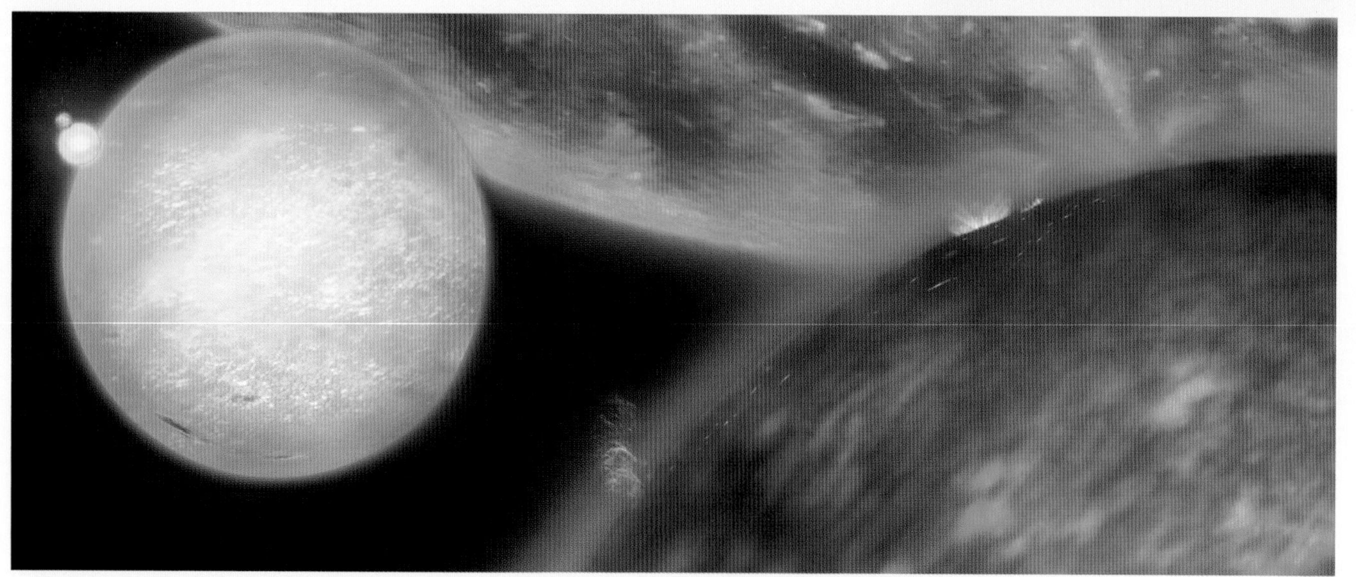

Above: *The view from a rocky exoplanet orbiting a blue giant star. These stars, like Rigel in Orion, have surface temperatures in the neighborhood of 9,725°C (17,540°F) and span diameters 5–10 times that of the Sun.*

Left: *A parade of stars. From left to right, the red dwarf Proxima Centauri, the Sun, the blue-white dwarf Sirius, blue giant Rigel, red hypergiant Mu Cephei (behind), and the red supergiant Betelgeuse. Of all these stars, the longest-lived will be Proxima Centauri, and the shortest will be the fast-burning Mu Cephei.*

F and A stars are slightly larger and warmer than the Sun, making up 4 percent of all stars. The gigantic B stars are even hotter and much rarer; they make up only one-tenth of one 1 percent of the main sequence stars. Finally, the huge O stars are the rarest of the rare. Some of these beauties shine several million times more brightly than the Sun, but they pay a price: their lives are short and end in violent explosions that result in white dwarfs, neutron stars, pulsars, or black holes.

Neutron stars have masses greater than the Sun, but are packed into spheres the size of a small city. As a star explodes in a supernova, its protons and electrons merge together, forming neutrons, and the explosion causes the stellar remnant to rotate at great speed. Some neutron stars spin 430,000 times each minute.

Perhaps the majority of main sequence stars play host to families of planets. It is tempting to think that the stars most similar to our own are the best places for our life search, but such may not be the case. Planets have left their signatures in the light of many stars, and those stars span much of the gambit of star types, from pulsars to red dwarfs to giants. Is it possible that biology has taken hold on worlds bathed in the radiation of a neutron star, or seared by the brilliance of a blue giant? Our search is just beginning, as is our understanding about the genesis and evolution of life.

A GIANT OF COSMIC PHILOSOPHY: A PROFILE OF GIORDANO BRUNO

One of the earliest to ponder life in the cosmos was the Dominican friar and philosopher Giordano Bruno. Born in 1548, Bruno grew up in a universe with the Earth at its center. The Copernican view of a Sun-centered solar system—with the Earth as merely one world circling it—was only beginning to influence European thinking. Most of Bruno's Catholic colleagues believed in a universe that circled around the Earth, with stars as points of light fixed on a crystalline sphere, a notion known as the Aristotelian view. But the scientific tides were turning and Bruno's contemporaries, including Galileo Galilei and Johannes Kepler, rose up to posit new ideas.

Bruno's concepts about the cosmos included an "endless and limitless universe." He snubbed the then-popular

Below: *Giordano Bruno's view that stars were distant suns seemed revolutionary to his contemporaries, as did his idea that other solar systems might have inhabited planets.*

THERE ARE ALSO NUMBERLESS EARTHS CIRCLING AROUND THEIR SUNS, NO WORSE AND NO LESS THAT THIS GLOBE OF OURS.

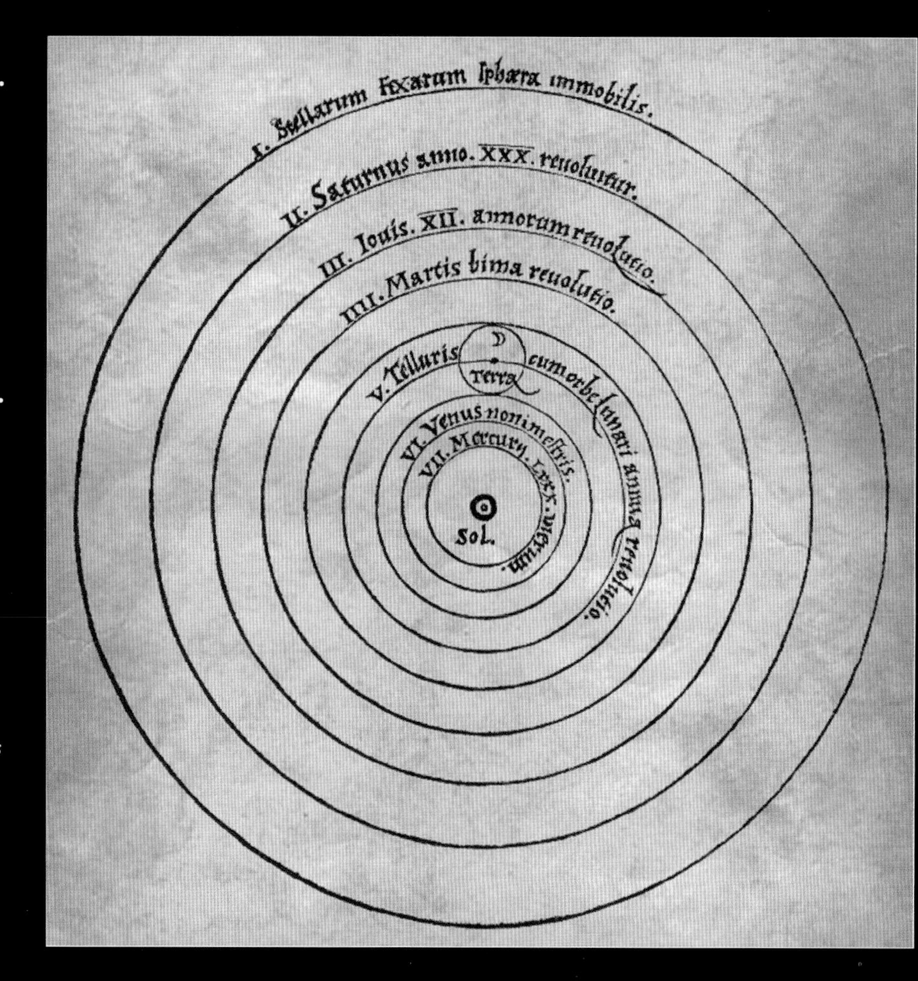

Right: *Bruno saw the universe not as a limited series of crystalline spheres surrounding Earth, but rather as "endless and limitless." In Bruno's time, the Copernican idea of a Sun-centered solar system was just gaining favor in the scientific community. Its social acceptance was years away.*

Below: A *bronze statue of Giordano Bruno in the Campo de' Fiori, Rome, the sculpture by Ettore Ferrari.*

idea of the planets affixed to spheres, claiming instead that they moved under their own inertia. In this, he was right. The planets of our solar system circle the Sun in roughly concentric paths. The entire planetary affair covers a lot of territory, so much so that astronomers have had to come up with a new yardstick for measuring it. This standard is called an Astronomical Unit (AU). An AU is the distance between the Earth and the Sun, 150 million km (93 million miles). The closest planet, Mercury, orbits the Sun at a distance of 0.4 AU, while Jupiter resides at 5.2 AU, and Neptune at over 30 AU. These numbers become important in the study and comparison of other planetary systems. Most significantly, Bruno suggested that the stars were distant suns, and that those suns had their own families of planets, perhaps worlds inhabited by beings not unlike us.

Bruno is sometimes held up as a martyr of modern science. But it is important to understand that the Catholic Church of Bruno's time did not oppose the Copernican, Sun-centered view of the cosmos. Bruno was executed in 1600 because of his views on God, the Trinity, and other theological points, and though he was personally unpopular, his death had nothing to do with his assertion that distant suns may harbor a multitude of worlds.

Bruno's view of the cosmos was uncannily prescient. The stars are, in fact, other suns, though many are quite different from the one gracing Earth's sky.

FINDING AN EXOPLANET IN A HAYSTACK

Claims of exoplanet discoveries stretch back to the 1800s, but none could be confirmed. In 1917, astronomer Adriaan van Maanen spotted a star while searching for a stellar companion to the nearby star Lalande 1299. Van Maanen's photographic plates held the first evidence of exoplanets: a faint ring of debris encircling the star. The ring, hidden within the old spectroscopic plates, wasn't discovered until 2016. While no planet was imaged, similar rings are usually associated with planetary systems.

In 1989, researchers finally discovered planets orbiting an unlikely star. Cornell University researcher Alex Wolszczan

Above: *Glowing with radiation, the dead planet Phobetor shimmers in the harsh environment near its pulsar sun. Any life there was sterilized into oblivion when the planet's sun collapsed into its superdense state.*

Right: *A neutron star, the collapsed core of a giant star but of small size and very high density, is here envisaged hurtling down towards continental Europe.*

IN THE TIME SINCE THESE EARLY DISCOVERIES, PROGRESS HAS BEEN MADE IN OUR OBSERVATIONAL TECHNIQUES AND INSTRUMENTS, AND THE TALLY FOR CONFIRMED EXOPLANETS IS RISING PRECIPITOUSLY.

used the giant Arecibo radio telescope not to find exoplanets, but to study his area of expertise: pulsars.

Pulsars are rapidly spinning neutron stars that spit out prodigious amounts of X-rays and other radiation, which usually shoot out in a direction other than through the poles, thus swinging across space in a great arc. Pulsars are the lighthouses of the galaxy: their beams of radiation sweep around many times each second as the star spins, making the star appear to blink on and off. For this reason, they were once thought to be signals from intelligent beings of an alien race.

As pulsars spin, they emit very regular pulses of energy, also many times each second, and these stellar heartbeats are so consistent that they can be used as cosmic clocks. Wolszczan was taking advantage of a rare opportunity: Arecibo's motors were under repair, so the paralyzed antenna could not move. The astronomer began to scrutinize the star PSR B1257+12, which happened to be in Arecibo's field of view. This pulsar rotated 161 times each second, but Wolszczan and his associate Dale Frail discovered that it was ill-behaved: its pulses varied on a 66.2-day cycle. Further analysis revealed another cycle lasting 98 days. The researchers announced an amazing discovery: planets were orbiting the pulsar and disturbing its pattern of spin. The inner alien world is 3.5 times the size of Earth, while the outer is slightly smaller.

PSR B1257+12 lies some 2,300 light-years away. Its trio of rocky planets have been named Draugr, Poltergeist, and Phobetor, and these planets make up a dead solar system, one that began with a fairly normal star that collapsed into a pulsar, blowing away the atmospheres—and perhaps some of the surface

Above: *A neutron star pulls gases from the surface of a red dwarf companion star. Many stellar pairings exist and if the stars are close enough, they will often exchange material, sometimes with explosive results. Gases accumulating on a star from a companion often result in a nova.*

Left: *Its pulsar sun collapsed to a globe 10km (6.2 miles) across, the planet Draugr simmers beneath a barrage of deadly radiation. Draugr is named after creatures in Norse mythology.*

material—of nearby worlds. PSR B1257+12 gushes deadly radiation, so the planets circling this star will be well sterilized and nothing like the Earth. Surface temperatures on the pulsar itself reach 28,600°C (51,500°F).

In 1993, another pulsar was found to be varying in its spin. A team led by Donald Backer were studying a double-star system, one a pulsar and the other a white dwarf. The pulsar, PSR B1620-26, was spinning 100 times every second. As with Alex Wolszczan's discovery a few years before, the regular pulses showed a telltale disturbance in their pattern. The two stars behaved as if they were being circled by a third star, but the astronomers soon discovered that the third body was too small to be a star. They announced the discovery of a planet there. PSR B1620-26 b orbits both the pulsar and the white dwarf at a distance of about 23 AU, or about the distance between the Sun and Uranus.

Left: *Gamma Cephei is a double-star system accompanied by at least one planet. The star Gamma Cephei A is a K-type orange star smaller than the Sun, while Gamma Cephei B is a red dwarf (seen at lower left). Gamma Cephei Ab is a giant planet with a mass at least 1.5 times that of Jupiter; it orbits the larger star.*

Right: *A black hole within its accretion disk would appear as the top view, if it were not for the effects of the warping of space and time around it. The image at bottom takes these bizarre shifts of light into account: the far side of the accretion disk is warped upward into an arc, while the opposite (underside) of the disk appears—as if from above—underneath the edge of the disk.*

The PSR B1620-26 system causes astronomers to wonder how planets got there in the first place. One scenario put forward is that one member of the star duo reached the end of its natural life (the main sequence) and exploded, leaving behind a slowly spinning neutron star. The companion star survived the explosion, and began to swell with age, becoming a red giant. Gases from its outer layers flowed onto the neutron star, adding to its spin, and as the gases formed a disk around both stars, planets formed within. Some of those planets survived as their aging sun became a white dwarf. Black holes are even more massive, and so dense that space and time actually warp around them. If a star is large enough, its core will continue to collapse as the star goes supernova, eventually shrinking to a point. This point is called a singularity. Around it, space and time are warped so that nothing—not even light—can escape. This region is called the event horizon. Often, a black hole is marked by a disk of dust that swirls around it, called an accretion disk.

Using the radial velocity technique (see page 27), Canadian observers proclaimed the discovery of a planet orbiting the primary star of the double-star system Gamma Cephei. Because of the star's variable brightness, many expressed doubts, but the planet was confirmed in 2002. Still, conventional planets around Sun-like stars were elusive. The Canadian magazine *SkyNews* summed up the situation in a 1995 article entitled "Where are all the Jupiters?" The article outlined efforts by Canadian and other teams to examine "21 nearby stars similar to our Sun. They are confident that their technique should show the gravitational signature of a Jupiter-size planet in an orbit approximately the size of Jupiter's orbit around the Sun. But they found none." Astronomers at the time didn't realize that those Jupiters were, in fact, out there, but they were hiding in orbits unexpectedly close to their stars.

In the time since these early discoveries, progress has been made in our observational techniques and instruments, and the tally for confirmed exoplanets is rising precipitously. As of July 2019, 4,096 exoplanets had been confirmed. In the summer of 2016, Researchers at the European Southern Observatory (ESO) confirmed planet Proxima b in the nearby Alpha Centauri system. And while another planet—Proxima c—appears to orbit farther away, the really great news is that the smaller Proxima b may be "Earthlike."

LIVING THE GOOD LIFE (IN THE GOLDILOCKS ZONE)

NASA Goddard Space Flight Center astrobiologist Ravi Kopparapu studies the habitability of exoplanets, and his work has led to insights into what makes a habitable zone, also called the "Goldilocks zone," habitable. When searching for an Earthlike planet, he says, researchers work through several steps. "Is it rocky? Is its distance from the star within range of lethal radiation?" Kopparapu and others are looking at planets of 0.5 to 1.5 Earth radii, the reasoning being that Mars measures about half an Earth radius, while planets half again as large as Earth are thought to retain terrestrial—or "terran"—natures. Then comes the habitable zone, Kopparapu says. He offers another way of thinking about habitable zones: "At the inner edge, liquid water condenses into clouds. At the outer edge you get condensation of CO_2."

The rationale for defining Goldilocks zones around other stars is to target selections for future planet-hunting observatories, such as NASA's James Webb Space Telescope and the European Space Agency's (ESA) Gaia. "We don't have the luxury—yet—of going to extrasolar planets to find their habitability," Kopparapu explains. "So what we can do now is remote observations. We try to target a planet that we think may be habitable, look at its atmosphere and see if it has any potential biosignatures, gas that may indicate life on the planet. The habitable zone is merely the first criterion to get to that point."

Below: *Hypothetical views from planets orbiting in the habitable zones of the red dwarf Lalande 21185 (left), the sun-like G star Kappa Ceti (center), and the hotter yellow-white F star Zeta Leonis (right). Although the distance to each sun varies, conditions on the surface of each planet remain similar. Lalande's habitable zone is much closer to the cool star than is the zone for the much hotter Zeta Leonis.*

Right: *Our inner solar system is superimposed on other types of stars to demonstrate habitable zones. In the center, the Sun, a G2-class star, is encircled by a habitable zone encompassing Venus, Earth, and Mars. Venus and Mars are on the inner and outer edge. Top: For an M-class star (a red dwarf), the habitable zone is much closer, as the star is small and cool. Planets within the habitable zone are often inside the distance of Mercury to our Sun, and our familiar terrestrial planets would circle well outside of it. But in the case of an F-class yellow star (bottom), brighter and hotter than the Sun, the habitable zone is considerably farther out.*

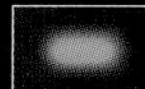 INDICATES HABITABLE ZONE

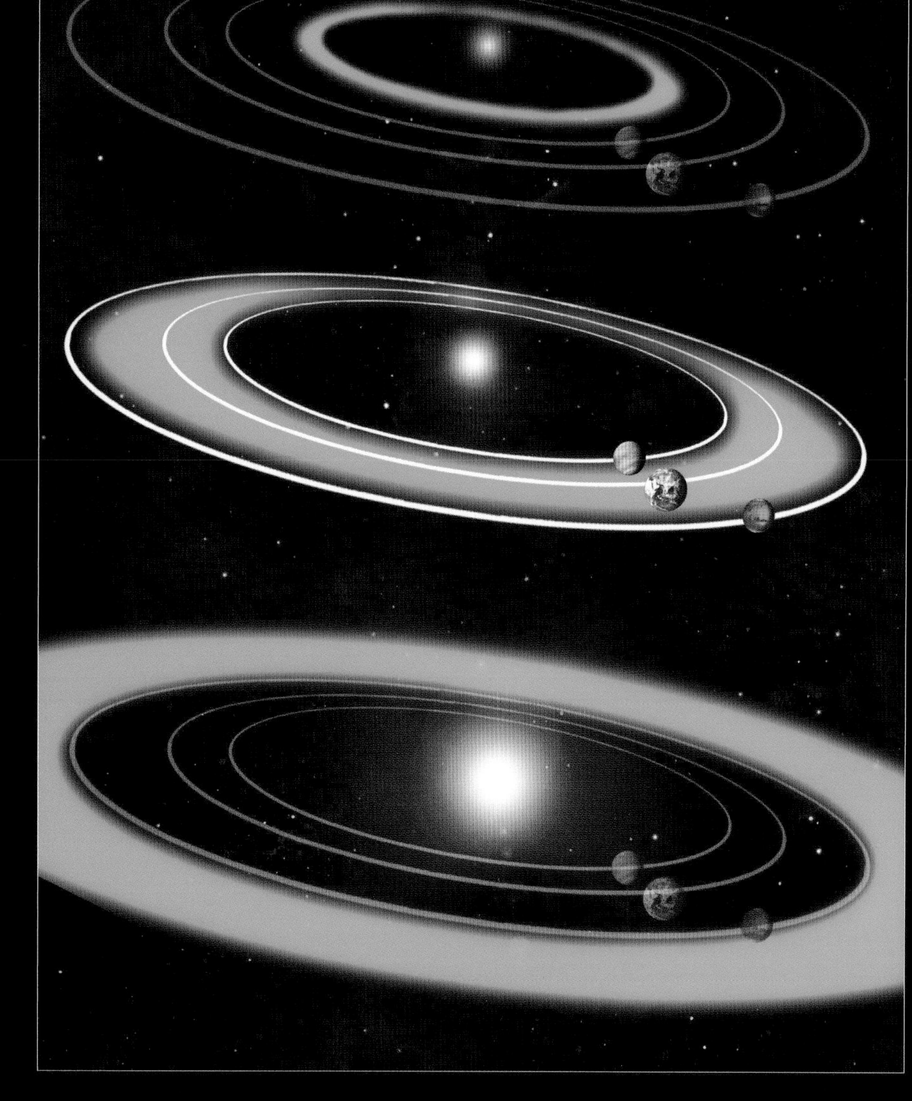

When astronomers ultimately detect a planet through the transit or radial velocity techniques, the first step is to find out how far the planet is located from its star. Once researchers have pinned down that important quotient, Kopparapu and his colleagues "can then use a climate model to see what would be the inner and outer edge of the habitable zone for the planet. Our models say, 'Within these stellar radiation levels, this planet of this size can maintain liquid water on the surface.' That is the range that defines the zone. The outer edge is too cold—snow covered, for example—and at the inner edge perhaps the planet is too close, so any water that exists may boil off completely. The habitable zone is the region around the star where surface liquid exists."

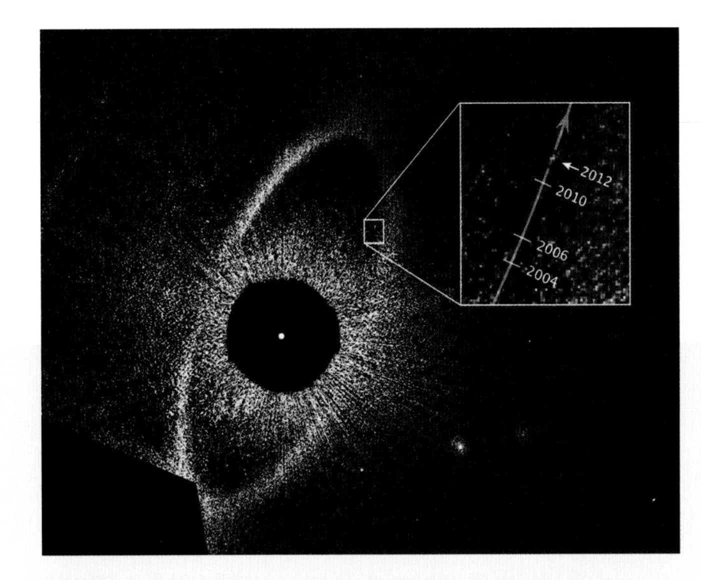

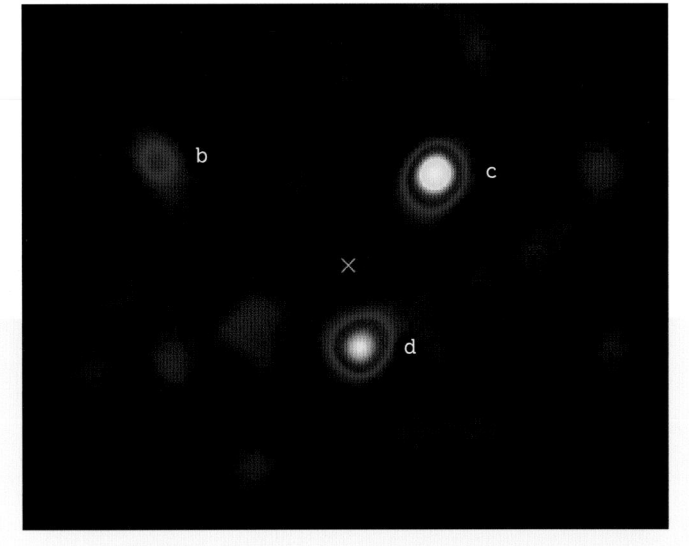

Above left: *With the light of the star Fomalhaut blocked out, the planet Fomalhaut b can be seen within a great disk surrounding the star.*

Above right: *The Palomar Observatory's Hale Telescope captured the glow of three planets circling the star HR8799 (located at the "X"). The star's light has been blocked out to reveal planets within its glare. The planetary trio—called HR8799b, c, and d—are thought to be gas giants like Jupiter, but more massive. While Jupiter orbits our Sun at about five times the Sun–Earth distance, these giants orbit their host star at 24, 38, and 68 times that gap.*

Astronomers have come up with several techniques for finding planets hidden within the light of a distant star. The first, and perhaps most difficult, is *direct imaging*. This technique entails imaging actual planets in orbit around other stars. Planets are far fainter than their host suns and are so close to the parent stars that they are lost in the glare. Gas giant–sized planets can be imaged more easily than the small, rocky terrans. To image distant planets directly, the glow from the planet's star must be blocked out, or "occulted"—the scattered starlight that blinds a telescope's view can be canceled out, revealing the planets orbiting close in. One promising technique for this glare removal is called interferometry, in which two ground observatories or space-based telescopes can be linked in phase so that their view of the sky becomes an interference pattern. With correct pointing, observers can target their telescopes so that the starlight falls in a minimum of the fringe pattern, essentially vanishing. The trick is to ensure that a planet is *not* in this cancelled-out zone, so that its faint light becomes visible.

Direct imaging requires the biggest, most powerful telescopes. Constructing large single mirrors is inefficient and expensive, but if engineers can craft several smaller telescopes and combine their light, the telescopes' power adds together to create higher resolution. In fact, a network of smaller telescopes chained together at a distance from each other resolves objects as if it were a single giant telescope covering the entire area. Some ground-based telescopes today are as powerful as orbiting ones like the Hubble Space Telescope.

Despite its challenges, the direct imaging technique has some distinct advantages over other approaches. A direct image provides astronomers with a spectrum, a chart of reflected light, from the planet itself. A spectrum hints at what the planet is made of, whether it has clouds, its temperature, and how it changes from one observation to another. These critical indications help us to understand these distant points of light as rich, diverse worlds.

It was more than a decade after the discovery of the first exoplanets before researchers were able to reveal the first direct images of exoplanets. At the time, over 300 exoplanets had been detected indirectly, using an assortment of exploratory techniques.

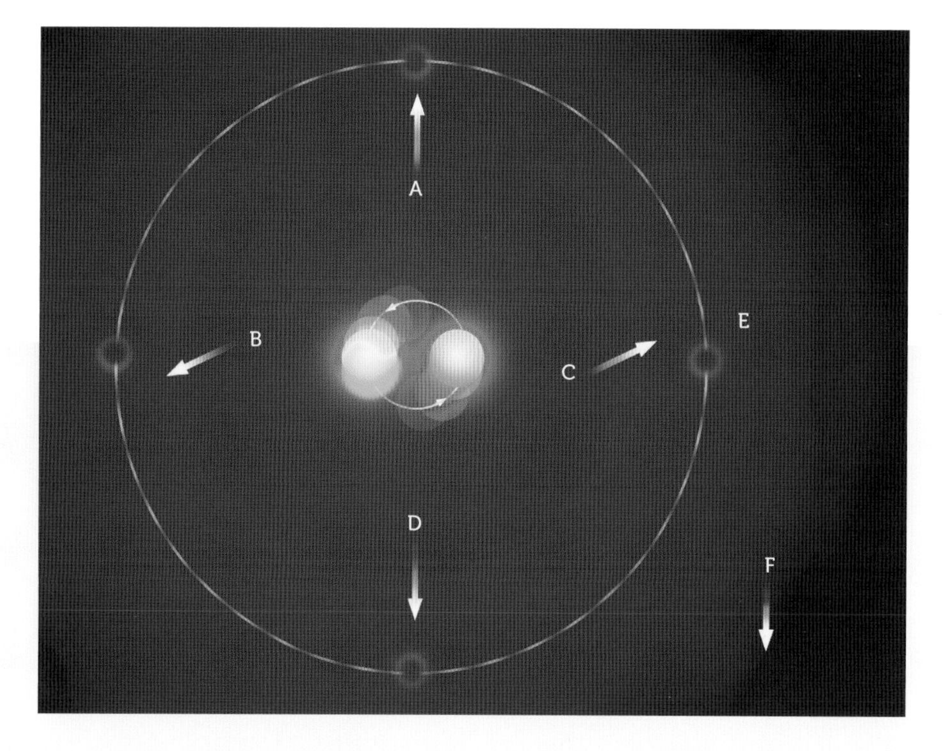

Right: *As an unseen planet pulls on its star, the star's light appears to shift toward blue as it moves slightly toward the Earth (B), and red as it moves away (C). This miniscule shift in the color of starlight reveals the presence of a nearby planet. When the planet is directly in front or behind the star (A and D), no change is detected. (F) shows the direction to the Earth.*

Most exoplanets are discovered by indirect means. One such approach in our search for Earth 2.0 is called the *radial velocity* method, which uses the spectrum of a star to detect movement as the gravity of unseen planets cause it to move back and forth. As a star moves, its light shifts due to the Doppler effect, just as the pitch of a horn changes with a passing car. As the star is pulled toward us by nearby planets, its light becomes bluer, and as its planets pull it away from us, the light becomes redder (it is "red-shifted"). In this way, subtle changes in a star's spectrum bear the imprint of the tug of nearby planets. Astronomers used this technique to discover the very first known exoplanets, including the massive Hot Jupiter 51 Pegasi b. The technique works best for large planets with heavy mass and was first tried in 1952; within a year 15,000 stars had been measured, but the instruments were not precise enough to detect the subtle changes in light. The best telescopes had errors of 1,000m/sec (3,280ft/sec) for the stars' movements, far too large to confirm nearby worlds. But new instruments came on line in the 1980s, and since then observers have been able to detect changes in a star's speed by as little as 7m/sec (23ft/sec). By comparison, Jupiter's influence on our Sun causes it to wobble at about 12m/sec (39ft/sec), so these detectors were good enough to find giant planets. Today, sensors like Chile's HARPS (High Accuracy Radial Velocity Planet Searcher) can sense changes as low as 0.3m/s (1ft/sec), accurate enough to find terran exoplanets. By 2016, roughly 30 percent of all exoplanet discoveries came from radial velocity studies.

Radial velocity has another advantage over other approaches: it can lock down the mass of a planet. Other techniques chart orbits and—sometimes—sizes, but radial velocity can tell how strong a pull the planet has on its star. The more mass a planet has, the more it will pull on its star; researchers can estimate a planet's mass by the amount of shift in its sun's starlight.

A DIRECT IMAGE PROVIDES ASTRONOMERS WITH A SPECTRUM, A CHART OF REFLECTED LIGHT, FROM THE PLANET ITSELF.

Another strategy for unveiling exoplanets is called *gravitational microlensing*. Microlensing takes advantage of a prediction in Albert Einstein's general theory of relativity. Einstein put forth the notion that mass distorts the space around it. A gas giant warps the space around it more than a smaller terrestrial planet does, and a star warps space more dramatically than a planet. Because of this phenomenon, light bends when it passes near a large mass, and this warping of space bends images behind a heavy object like a star.

The effect was confirmed by Sir Arthur Stanley Eddington and a team of astronomers during a total eclipse in 1919. By tracking stars as they disappeared and reemerged from behind the Sun during totality when the Sun's light was blocked, the observers demonstrated the bending of starlight around the Sun. Concurrent observations were carried out in in Sobral (Brazil), and in São Tomé and Príncipe, islands in the Gulf of Guinea off the west coast of Africa.

Luckily for Earth 2.0 hunters, the same phenomenon can be applied to planets circling stars. When one star appears to pass near another (from the viewpoint of Earth), the closer star acts as a lens, increasing the brightness of the farther star. But if a planet is present in orbit around the closer star, a second brightening of the far star will occur. These brightening events are called microlensing or binary lens events. Microlensing events are brief because they must occur when two stars are almost perfectly aligned. And they unfold quickly—over days or weeks—because the two stars and the Earth are all moving in space, and moving in relation to each other. A thousand of these events have been observed in the past decade. By early 2011, 11 exoplanets had been found using this technique, one of which was, up to that point, the least-massive exoplanet discovered around a main sequence star. The NASA Exoplanet Archive lists 76 confirmed discoveries using this technique.

Microlensing has several advantages over other approaches. First, it can detect planets at great distances. Other techniques are most effective within 100 light-years of Earth, but gravitational microlensing has actually spotted planets near the center of our galaxy, 26,000 light-years away. Another important advantage of the microlensing method is that it can find planets orbiting at great

Above: *The eclipse of 1919 provided Arthur Stanley Eddington (left) with an opportunity to see the Sun's gravity deflecting the light of stars during totality, when the Moon covered the disk of the Sun. Eddington's team was stationed on the African island of Príncipe, while a second team observed at the same time from the town of Sobral in Brazil (center). Eddington's observations confirmed part of Einstein's theory of relativity, but also demonstrated one of the first practical uses of gravitational microlensing.*

THE KEPLER SPACE TELESCOPE SIMPLY STARED AT ROUGHLY 145,000 STARS AND SEARCHED FOR CHANGES IN LIGHT LEVELS.

distances from *their* stars. While other techniques are best at finding planets in close orbits, microlensing can fill in the gaps of worlds orbiting the outskirts of their solar systems. And finally, microlensing covers great swaths of the sky, able to search tens of thousands of stars at once.

Then came the Kepler Observatory. Its transit technique proved to be the most fruitful of any yet. The transit approach to planetary hunting is gloriously straightforward: the Kepler Space Telescope simply stared at roughly 145,000 stars and searched for changes in light levels. When carefully monitoring the brightness of a star, any drop in that brightness might betray the passing of a planet in front of it. Multiple crossings provide information about the planet's orbit and even mass (see "The science of transits" feature on page 31 for more information).

A related planet-searching method is called the *transit timing variation* (TTV) technique. When a planet is revealed by the transit technique, additional observations of its passes sometimes

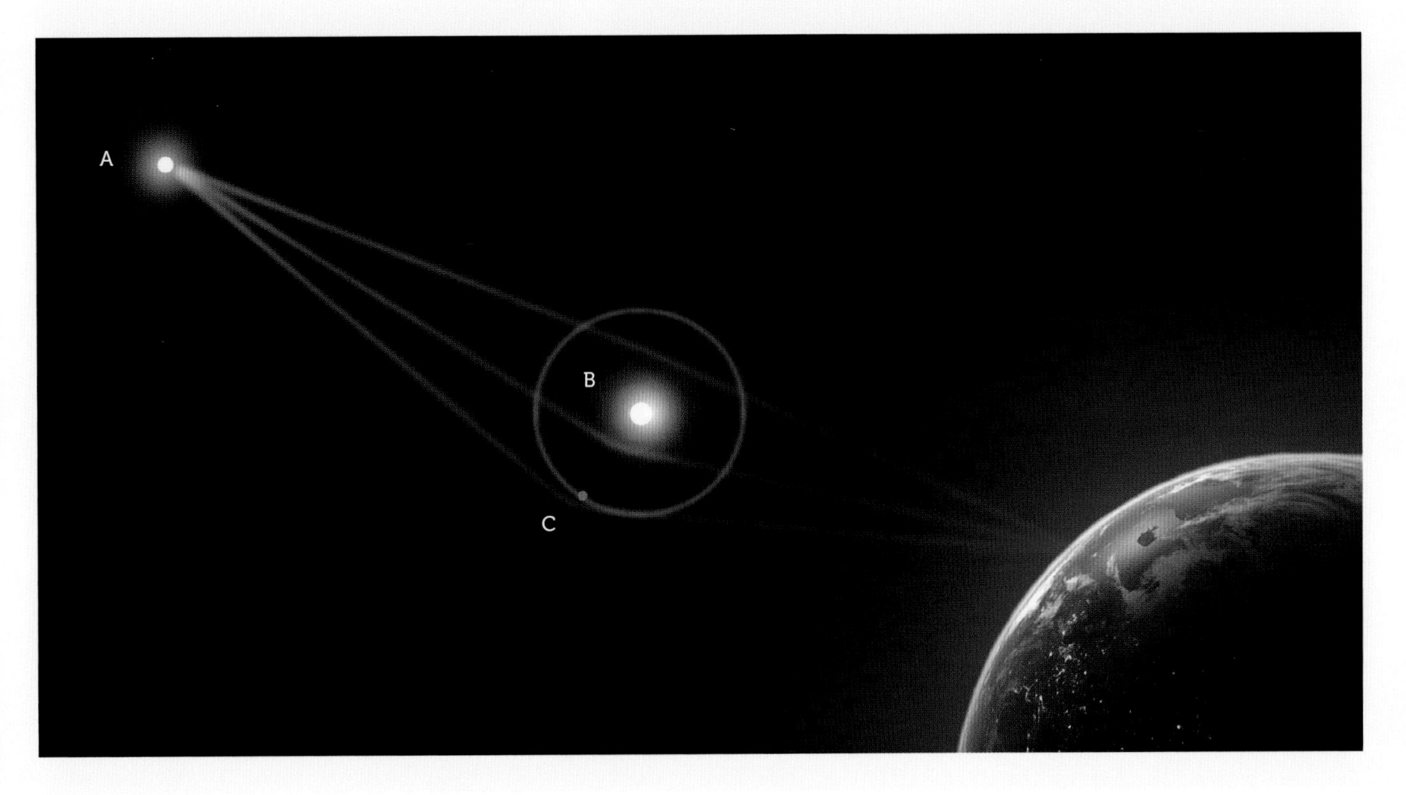

Below: *Two stars are involved in microlensing. A distant star (A) acts as the source light, while the closer star (B) acts as a "lens"; its gravity bends space around it and the starlight shining through. If the closer star has a planet (C), that planet's own gravity adds to the lensing effect.*

AN OBSERVER IN A NEARBY STAR SYSTEM COULD EASILY SEE THE EFFECTS OF JUPITER'S PULL ON OUR SUN.

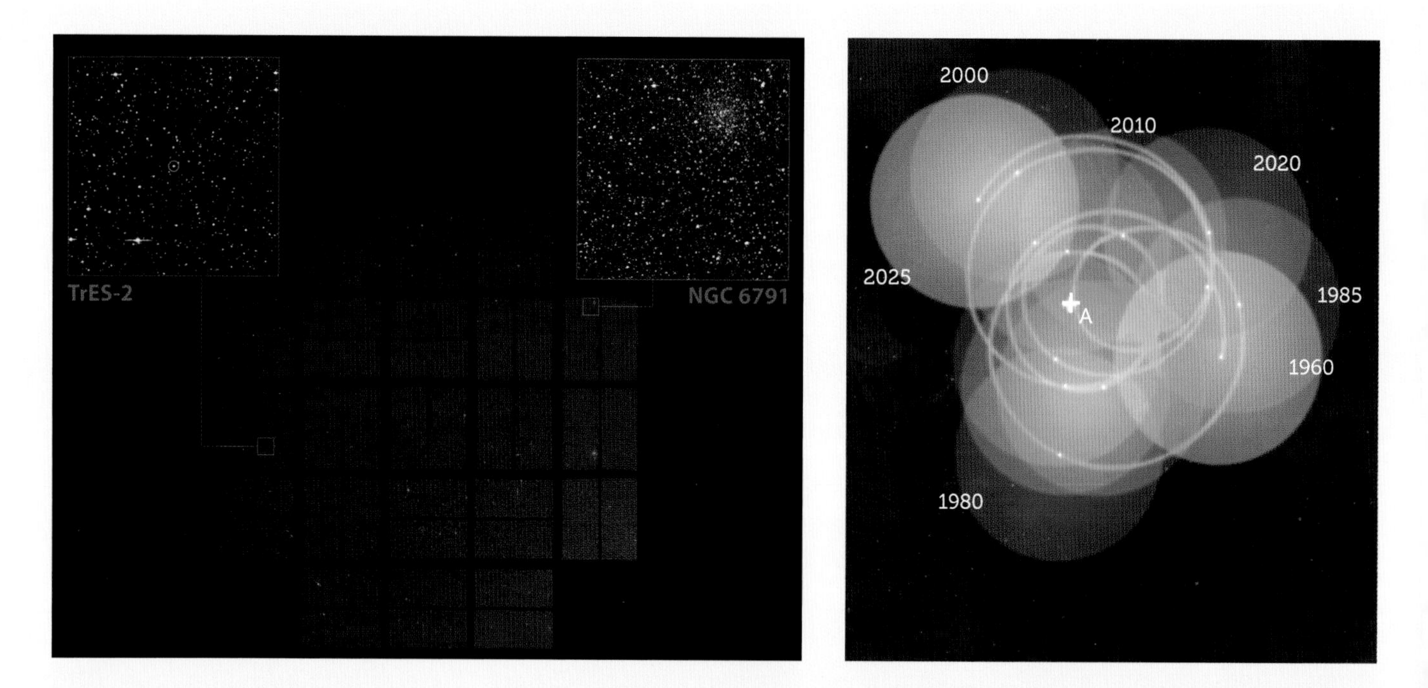

reveal variations in the *timing* of its transits. These variations can be due to other nearby planets that are not lined up to transit the star. This method is so sensitive that it can locate smaller planets with Earthlike masses.

Astronomers are honing other planet-hunting tools. While their radial velocity approach uses changes in the color of starlight to sense a star's movement, *astrometry* charts the position and shift of a star against the background starfield. Astrometry can reveal the inclination of the planet's orbit, which helps to determine its mass. The mass helps observers to track down its bulk composition and surface gravity, which in turn discloses something about the nature of the planet itself: Is it a low-density gas giant? Is it a rocky terrestrial planet? Is it likely to have an atmosphere or ocean?

Although the technique has yet to discover exoplanets, astrometry has confirmed several exoplanets previously discovered by radial velocity. But in our perennial search across the universe for Earth 2.0, our astrometry instruments will need to be hundreds of times as sensitive as they currently are. ESA's Gaia mission is carrying out an all-sky survey of a billion stars, using astrometry as one of its primary tools to locate and identify exoplanets.

Above left: *The Kepler Space Telescope was designed to stare at a small section of the sky, searching 145,000 stars for companion planets. Kepler's 21 charge-coupled device modules monitor five square degrees of the sky for changes in starlight levels. This image details two landmarks in Kepler's field of view: the star cluster NGC 6791 and the star TrES-2, where one of the darkest planets on record resides.*

Above right: *An observer in a nearby star system could easily see the effects of Jupiter's pull on our Sun. The diagram shows the relative position of the Sun against a fixed background (the lines indicate the Sun's central path; A=the center of mass). The Sun's center points are marked with each labeled year.*

THE SCIENCE OF TRANSITS. HOW DO THEY REVEAL EXOPLANETS?

The transit technique, also called photometry, was first described by Dutch astronomer Christiaan Huygens in a book called *Cosmotheoros*, published in 1698. Huygens was out to estimate the distance between the stars, using their brightness as a yardstick. Huygens's experiment, while flawed, was the first to make use of photometry. But centuries later, the technique was used by the Kepler Observatory, with spectacular results.

For Kepler, it's an exacting method; light levels drop by only 0.01–0.1 percent, depending on the size of the planet in comparison to its star. A transit takes place once in each orbit of the planet. The duration of the transit can reveal a host of information. As the planet passes in front of its sun, astronomers can discern the planet's distance from the star and orbital speed. The total amount of blocked light reveals the size of the planet. When combined with the radial velocity method (which determines the planet's mass by how much it tugs on its star), observers can glean the density of the planet. A planet's density speaks volumes about its physical structure. The planets that have been studied by both methods are the best understood of all known exoplanets, but they are rare.

Below and right: *An idealized view (below) of a planet transiting its star, with light curve below. (A=transiting planet; B=star; C=light level; D=time; E=light level from transit.) At right, the reality is not so neat, as seen in this view of a K2 sampling of transit light readings. Transits like these have much to tell scientists about a planet's size, orbit, and sometimes even mass.*

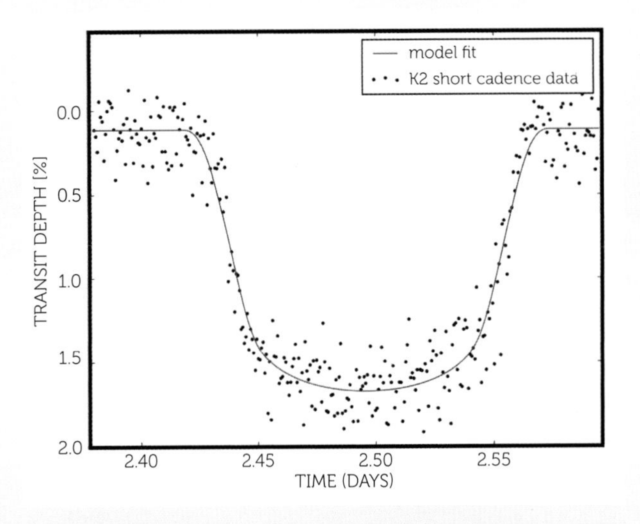

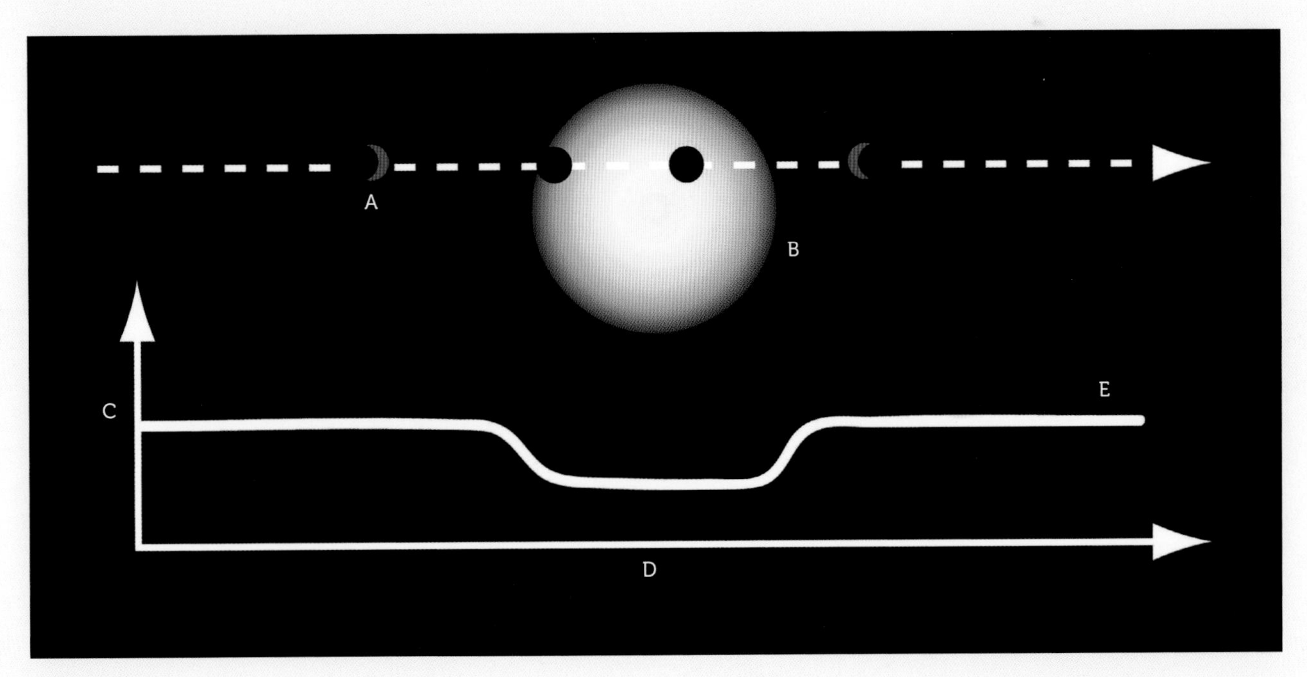

THE SEARCH BEGINS!

By the 1990s, planet hunters had found a handful of exoplanets, but none in orbit around a Sun-like star, and none as small as the Earth. They bagged their first catch at the main sequence star 51 Pegasi. Michel Mayor and Didier Queloz at the University of Geneva announced their discovery on October 6, 1995. They used radial velocity, the shift of starlight, to find the planet.

It had been a long search, as the team studied the spectrum of 150 Sun-like stars, watching for telltale signs of a wobble caused by nearby planets. With an official label of 51 Pegasi b, the discovery was a revelation, as the planet was huge—more similar in size to Neptune than Earth—but in a tight orbit close to its star. The planet orbits once each four days and is so close that surface temperatures are hot enough to melt copper (1,000°C /1,800°F). Another exciting aspect of the discovery was that the planet orbited a star very much like our own, the first such planet found.

Early the next year, more detections quickly followed the 51 Pegasi revelation. The observing duo of R. Paul Butler and Geoffrey Marcy unveiled the discovery of giants orbiting three other stars, 47 Ursae Majoris, 70 Virginis, and 55 Cancri.

47 Ursae Majoris b tips the scale at 3.5 times the mass of Jupiter, a behemoth of a gas giant about twice the distance from its star as Earth is from the Sun. Its path follows an orbit similar to a circle between Mars and the asteroid belt. Temperatures at its cloud tops are chilly, probably dipping to about –80°C (–110°F) even at high noon.

For its part, 70 Virginis b orbits the Sun-like 70 Virginis at a distance where conditions are fairly temperate. The planet is massive, weighing in at over eight times the mass of Jupiter, but temperatures at its cloud tops probably average 85°C (185°F). *Astronomy* magazine pointed out that this is the temperature of "lukewarm coffee." The giant world is probably a sterile, windblown cloudscape with no life, but a planet its size likely has large moons, and those natural satellites may well have atmospheres that moderate temperatures.

A third early discovery (in 1996), adding to the Jupiter-class exoplanets, was 55 Cancri b. The system of 55 Cancri is a double star within 41 light-years of Earth. 55 Cancri A is a K-type star that lies between the size of a red dwarf and a Sun-like G star. (55 Cancri B is a red dwarf.) Its first discovered planet, 55 Cancri b, is a Hot Jupiter orbiting its sun every 14.5 days. Another planet, 55 Cancri e, is a Super Earth racing around the star every 18 hours. It is the innermost known planet and was discovered in 2004.

By spring of the next year, astronomers had found more than 10 planets around Sun-like stars, but the radial velocity approach had its limitations, and most of its findings involved Hot Jupiters or other planets large enough to pull strongly on their stars. Still, among that family of oversized planets, many intriguing worlds are on display, with more than one-third

Right: *Four solar systems compared: the stars 51 Pegasi, 70 Virginis, and 47 Ursae Majoris are all G stars, similar in diameter to our Sun, but their major planetary companions orbit at varying distances. Here, distance is roughly to scale, with blue demarcations for 1 AU (the distance between the Earth and Sun). Planets are not to scale.*

THE DISCOVERY WAS A REVELATION, AS THE PLANET WAS HUGE—MORE SIMILAR IN SIZE TO NEPTUNE THAN EARTH—BUT IN A TIGHT ORBIT CLOSE TO ITS STAR.

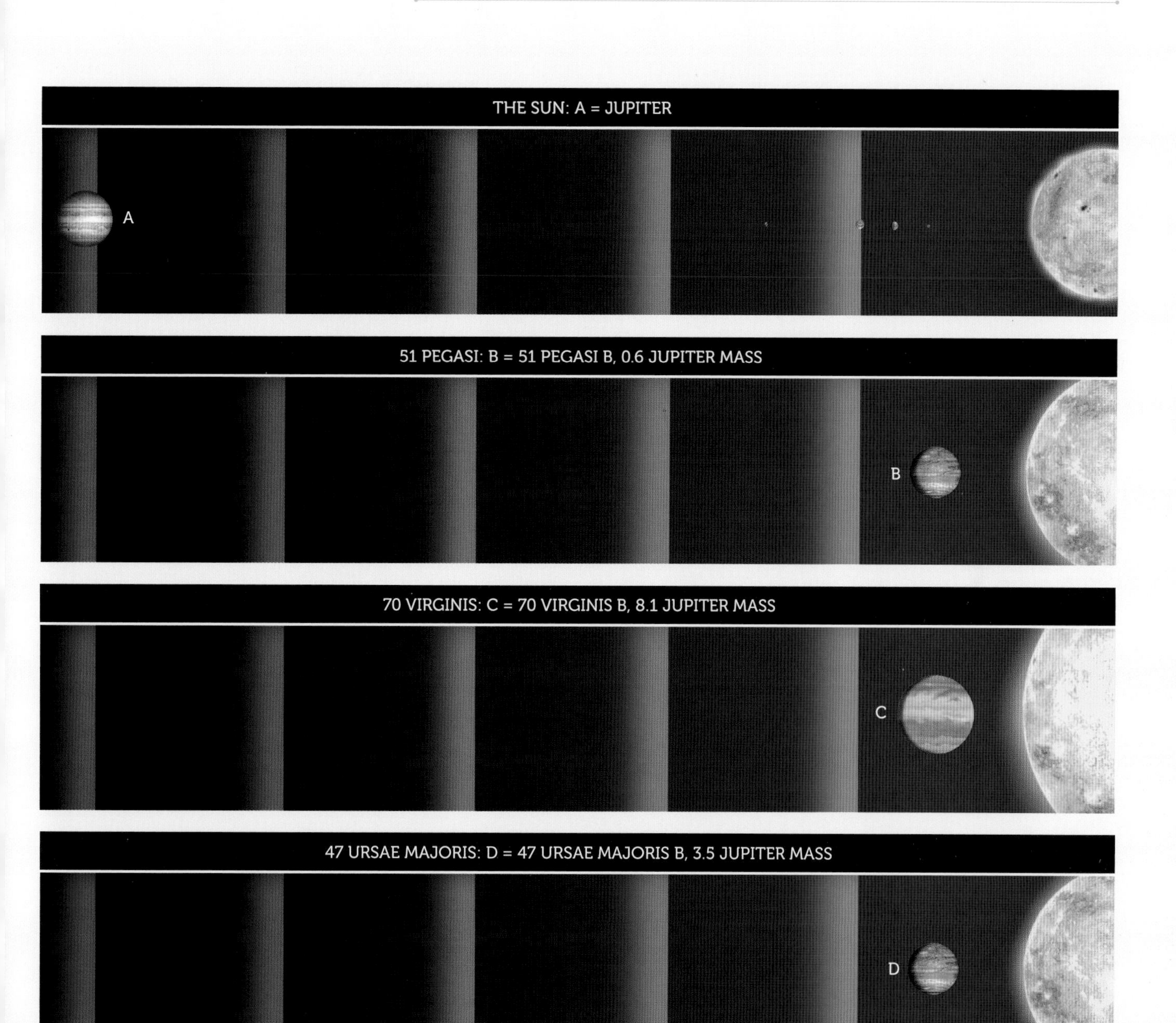

HOW DO EXOPLANETS GET THEIR STRANGE NAMES?

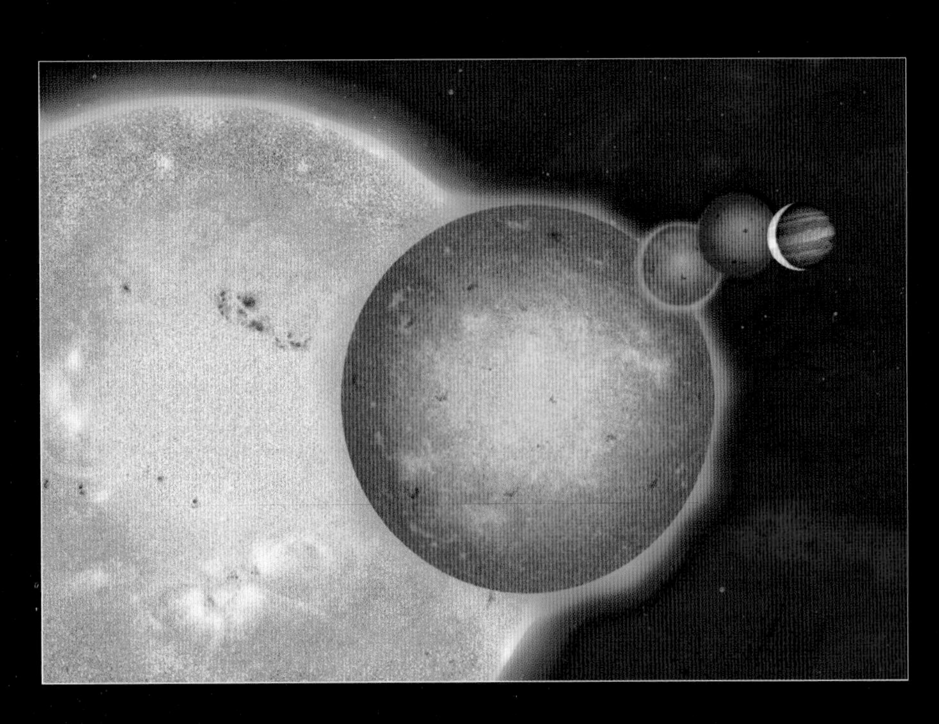

Exoplanets seem to have baffling designations to the uninitiated, names that are usually associated with letters and numbers in seemingly random arrangements. How does a planetary body get a name like HD 189733 b, for example? Or what about 51 Pegasi b? Gliese 581 c?

In general, an exoplanet is named after the star it orbits. It takes the name of the star and adds a lower-case letter. The first planet discovered receives a b (the a is assumed to be the star itself). The second planet is given a c, the third a d, and so on. If more than one planet is discovered at one time, the letters ascend with distance to the star, so that planet b is closer than planet c. This is the convention used in the planetary system of TRAPPIST-1, whose seven Earth-size

Above left: *A comparison of stars falling between the size of the Sun and Jupiter shows a wide variety of names. Each star also has its own character, with varying temperatures, roughly by size. The Sun (at left) is the hottest of these stars, with temperatures at about 5,525°C (9,975°F). Dwarf stars are cooler, like the Jupiter-sized WISE 1828, which simmers at about 125°C (260°F). In between are the large red dwarf Gliese 229A and the smaller, cooler Gliese 229B. Jupiter is shown for scale.*

Left: *Although closer to its star than the gas giant Kepler-20 b, this little world gets its name because it was discovered later. Kepler-20 e was the first planet smaller than Earth to be found orbiting a Sun-like star.*

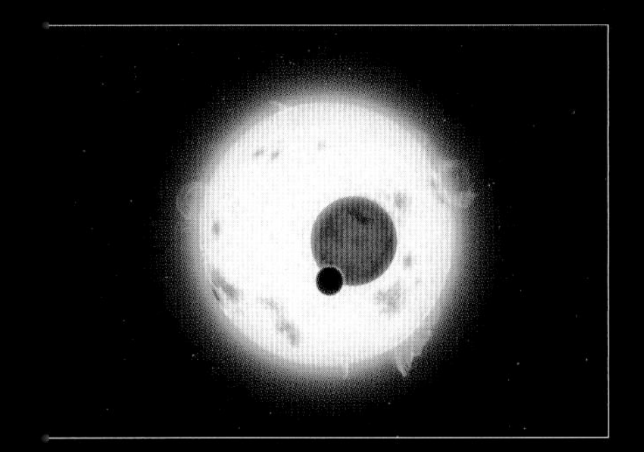

Above: *The planet seen here (in silhouette) has the ungainly name of Kepler-16 AB b. It orbits not one, but two suns. The largest of the two stars, an orange dwarf (or K star) called Kepler-16A, is about 70 percent the mass of our Sun, and the smallest, a red dwarf called Kepler-16B, is about 20 percent the Sun's mass. The two stars are aligned so that they pass in front of one another, while Kepler-16 AB b passes in front of both, blocking out a small portion of their light. Because it circles outside of both stars (which, in turn, circle each other), it is given its long train of letters.*

planets ascend from TRAPPIST-1 b to TRAPPIST-1 h (see Chapter 3 for details on this remarkable system).

The naming conventions become a bit trickier when the planet orbits within a multiple-star system. The International Astronomical Union (IAU), an international association of astronomers, names stars in order of their brightness. So, for example, in the triple-star system of Gliese 667, Gliese 667A is the brightest of the three, while Gliese 667C is the dimmest. The first planet to be discovered orbiting Gliese 667C is called Gliese 667C b, while the second planet found there is Gliese 667C c.

Some exoplanets have proper names. One example is 51 Pegasi b, a Hot Jupiter orbiting the Sun-like star 51 Pegasi. The planet was unofficially known for some time as Bellerophon, but was later officially named Dimidium, a Latin word meaning "half"—the name refers to the planet's mass, about half that of Jupiter.

IN GENERAL, AN EXOPLANET IS NAMED AFTER THE STAR IT ORBITS. IT TAKES THE NAME OF THE STAR AND ADDS A LOWER-CASE LETTER.

Above: *The Gliese 667C system (compared to Earth on the right) includes a possible three planets in the habitable zone, named by the convention of letters in order of discovery: planet c is the innermost, next out is f, while e is on the outskirts of the habitable zone.*

Below: *51 Pegasi b, now known as Dimidium, was one of the first exoplanets to receive an official name. This painting was done just months after the planet's discovery and was one of the first to show a Hot Jupiter, a then-new type of planet.*

of them falling into the Neptune-sized or Super Earth scale.

Discoveries continued to dribble in until the launch of the Kepler Space Telescope. The observatory dramatically increased the number and rate of discoveries of exoplanets. New telescopes, both on the ground and in space, add to the tally nearly daily. Observers using the next-generation follower to Kepler, NASA's Transiting Exoplanet Survey Satellite (TESS), have recently discovered a variation on Hot Jupiters, a type of planet bridging gas giants and rocky terrans. Like Hot Jupiters, Hot Neptunes orbit their stars closely but are smaller in size than their larger cousins. One of them is LTT 9779b, which orbits a star similar to the Sun, but at a scant 2.5 million km (1.6 million miles) away—23 times as close as Mercury is to the Sun. The searing planet races around its sun rapidly, once every 19 hours. Temperatures on the Hot Neptune probably reach a scorching 1,725°C (3,140°F).

The planet may be at a stage of evolution between a gas giant and a rocky terran world. Theories suggest that planets are slippery things, migrating from their place of birth in their solar

Left: *In a distant future, the last vestiges of atmosphere are stripped away from the remnant core of the Hot Neptune LTT 9779b. The roasted gaseous giant may be on track to becoming a dead, airless terran planet because of its proximity to its star.*

JUPITER'S GUTS ARE AN AMORPHOUS MIST OF HEAVY ELEMENTS PERHAPS SPANNING HALF THE PLANET'S WIDTH.

system to a different location. The "Grand Tack" theory describes the gas giants of our own solar system as having formed far from the Sun. As the solar system matured, gas and ice giants migrated inward, scattering asteroids, comets, and planets in their wake. Jupiter may have qualified as a Hot Jupiter for a short while, but as Saturn moved back outward, its gravity pulled Jupiter out with it. At the same time, the orbits of Uranus and Neptune were also shifting. If this cosmic shuffleboard is common among planetary systems, it may well be that LTT 9779b came in from the cold, spiraling inward from the cool outer region where it first formed. Migrating planets described by computer scenarios like the "Grand Tack" or the "Nice Theory" (named after the city in France) may explain the plethora of gas giants in tight orbits around their stars.

Today, LTT 9779b is in transition, and unless something dramatic happens to it, it will gradually lose its atmosphere. All that will remain will be its burned-out core, a ball of rock and metal devoid of air and water.

Initially, Hot Jupiters made up the majority of discoveries, simply because they were the easiest to find. But as the exoplanets added up, it became clear that while Hot Jupiters are ubiquitous throughout the galaxy, they are also strange. NASA Goddard's Eric Lopez has been modeling atmospheres of exoplanets, and he is intrigued by the extreme environments they present. "These Hot Jupiters hover at 2,000°C [3,632°F]. They have got winds whipping around at a thousand miles an hour." One puzzling aspect of these giants is their diffuse cores, something unexpected by planetary models. Astronomers expected giant planets to have massive cores, but such is not the case.

A recent study may shed light on why so many of these lightly compacted gas worlds orbit the stars we've studied. The Juno spacecraft has been mapping Jupiter's interior by investigating its gravity fields, and it turns out that Jupiter's core is less dense than models predicted. In fact, it is quite fluffy compared to the packed rock and ice core expected by planetary evolutionists. The current understanding is that Jupiter and other gas giants began as a compressed "seed" of rock and metal, slowly gathering ice and gases around it. But Juno's readings indicate that Jupiter's guts are an amorphous mist of heavy elements perhaps spanning half the planet's width. What could have led to such a structure,

Right: *A Hot Jupiter orbits within the exotic triple-star system of HD 188753. The main star is a Sun-like G star. A duo of smaller stars orbit each other and, in turn, circle around the main star at a distance of over 12 AU, taking 27 years to complete their yearly journey. Within this wild conglomeration, a Hot Jupiter was reported to circle the main star at a distance of only 8 million km (5 million miles), making the circuit every 80 hours. Here, planet HD 188753 Ab is seen from a moon.*

and why is it apparently so common among exoplanets? Recent computer simulations carried out at Houston's Rice University demonstrate that a massive pileup between Jupiter and another large planet could have fractured Jupiter's original dense core, mixing material into the outer layers of the planet. The studies show that traces of this scramble could have survived for 4 billion years, sustaining the mixed interior into the present. It would have taken a planet 10 times the mass of the Earth, however, to shatter Jupiter's core—which at that time would have been only 15 percent of the planet's diameter—and mix its heavy elements into Jupiter's upper layers.

There are rival scenarios to describe what may have led to Jupiter's diffuse core, and what may be happening within the cores of other exoplanet giants. One concept is that heavy elements mixed with lighter gases during Jupiter's formation. Another idea posits that strong currents dredged up material from deep within Jupiter's core, churning heavy elements into the upper layers of the planet. Researchers hope that with advanced computer modeling and further study of hot gas giants orbiting other stars, a likely answer will be found. For now, we continue to scrutinize the worlds around distant suns.

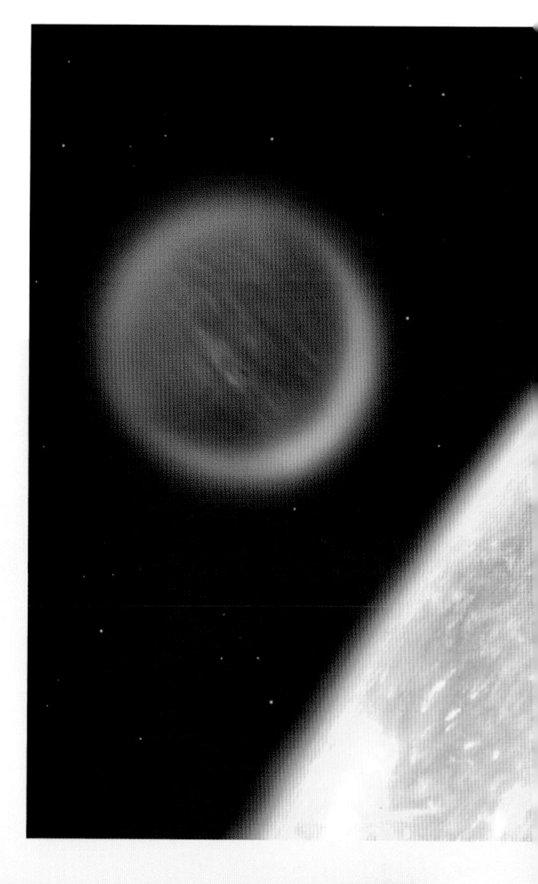

Left: *Another Hot Jupiter circles the F-type star WASP-142. The planet is half again as far across as Jupiter, but only eight-tenths as massive. This low-density world hugs its star, completing an orbit every two days. The giant planet is seen here from a nearby moon, whose rocks glow red-hot because of its nearness to the star.*

Right: *Some Hot Jupiters are even less dense—and probably contain more diffuse cores—than Jupiter or Saturn. These planets are "fluffed up" by the heat of their nearby suns, extending their atmospheres into amorphous globes.*

Below: *The Juno spacecraft, in orbit around Jupiter, is returning spectacular images of the planet's clouds. But the craft's main area of study is deep within, where Jupiter's magnetosphere and gravity are generated. Juno revealed that Jupiter has a remarkably dispersed core.*

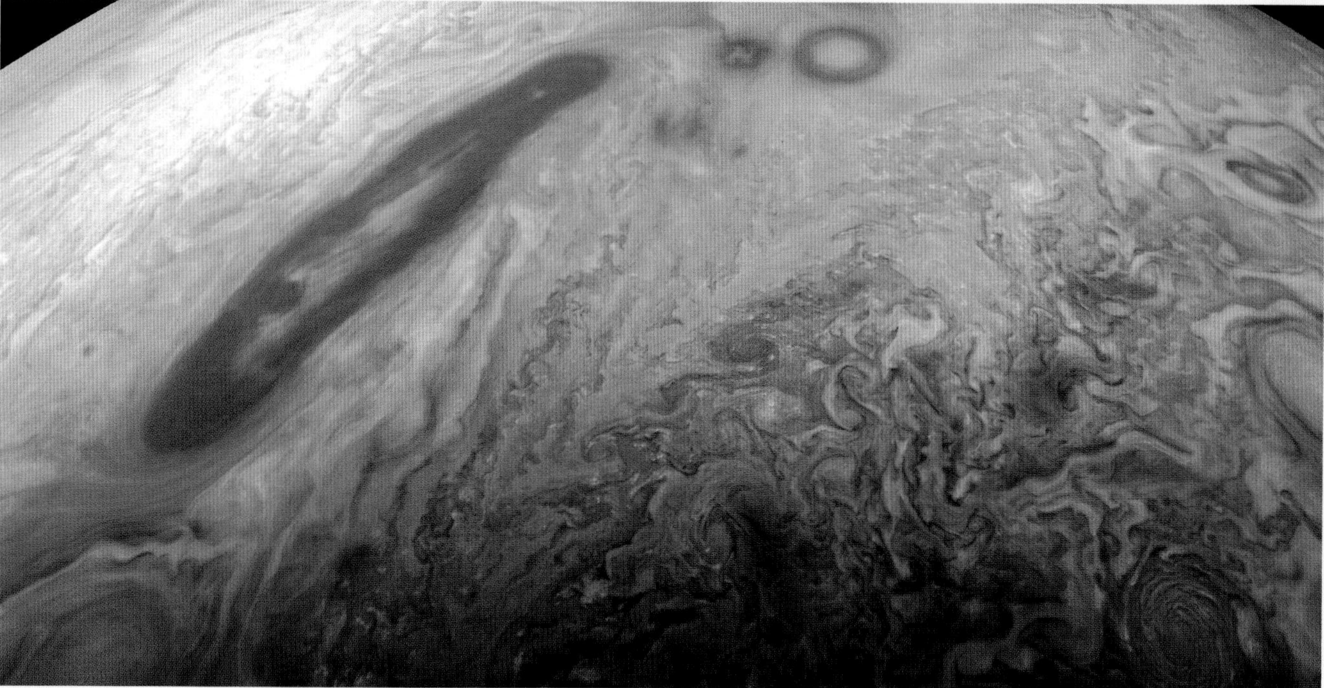

BEYOND THE HOT BALLS OF GAS

With all of those giant planets floating around, astronomers began to wonder about large moons. If our own solar system is any indication, large planets tend to have large families of moons. And in the rough-and-tumble evolution of a planetary system, some of those moons may well have been pulled from their home planets, cast away into other orbits. Small planets in systems with large gas giants may once have been moons of the larger worlds, now orbiting their star in their own path. These escaped moons have been nicknamed "ploonets."

Computer models indicate that as gas giants migrate from the outer system to the close orbits occupied by Hot Jupiters today, gravitational forces interact with their moons. The combined tug of the gas giant and the nearby star injects energy into the moon's flight path, gradually pushing it farther away from its planet until it finally escapes. The nomad planet then settles into its own orbit. Authors of a recent study suggest that any planetary system containing Hot Jupiters should have ploonets in orbit as well. Scientists may be able to detect ploonets by their effect on their former home planet: ploonet gravity will affect the speed of the larger planet as it transits across the face of its star. Ploonets may not be long for this world: nearly half of

Right: *A former ice moon drifts far from home. Some ploonets will be indistinguishable from planets, though most will have eccentric orbits.*

SMALL PLANETS IN SYSTEMS WITH LARGE GAS GIANTS MAY ONCE HAVE BEEN MOONS OF THE LARGER WORLDS, NOW ORBITING THEIR STAR IN THEIR OWN PATH.

A NEW WORLD

Astronomers have been studying the star CVSO 30, where they have discovered an exoplanet using the transit technique. Researchers have now combined the formidable capabilities of the European Southern Observatory (ESO)'s Very Large Telescope (VLT), the Keck Observatory atop Mauna Kea in Hawai'i, and the Calar Alto Observatory in Spain to find a second planet using direct imaging. This image was produced by the VLT. The new world, CVSO 30 c, is the tiny brown dot to the upper left of the bright star at center. The planet discovered earlier, CVSO 30 b, is a Hot Jupiter lost in the glare of the star in this view. CVSO 30 c orbits some 660 AU away, taking a leisurely 27,000 years to complete its course. If the planet is confirmed, CVSO will become the first system to have planets detected by both the transit and the direct imaging methods.

Above: *A possible ploonet 1,200 light-years from Earth.*

Left: *In the aftermath of Neptune's capture of Triton, gases from Neptune form rings around the planet. In the foreground, a major ice moon is on a path to depart the system, carrying some of the gas and debris with it. Encounters like this give rise to ploonets.*

Right: *Stripped from their home planets, some ploonets may leave their solar systems completely, wandering interstellar space as lonely cosmic vagabonds. Most will lose their atmospheres in the process, leaving the surface open to bombardment by meteors and comets. Cratered surfaces may be common among ploonets.*

the ploonets in the computer models either collided with their home planet or with their sun within the span of half a million years. Of the ones that are left, any in stable orbits might be indistinguishable from planets that formed independently. These will be rare. For the most part, ploonets will remain in unstable orbits, and if they are out there, observers hope to find ploonets in their native environment, circling precariously in their eccentric orbits.

Our own solar system has an example of ploonets in reverse. Neptune's moon

Triton likely formed in the outer solar system as part of the Kuiper Belt, and as Neptune migrated during early solar system formation, Triton was detoured from its path and captured into Neptune's gravitational field, where it orbits today. We know the massive moon came from somewhere else as its orbital direction is opposite to what it would be had it formed in the Neptune system. Additionally, Neptune itself is missing the kind of midsize moon attendants that the other gas and ice giants have, suggesting that many of its larger moons were ejected as Triton plowed into the system.

Left: *Today, Triton bears the scars of its encounter with Neptune and its moons. Its entire facade has been resurfaced by titanic forces, melted and refrozen into the bizarre landscapes we see today. Triton's surface has rocky outcrops, ridges, troughs, mesas, and dark-spotted plains, but few craters and little topographic relief. Dark plumes of nitrogen gas and dust erupt through the thin nitrogen ice crust, reaching up to 8km (5 miles) high before being blown long distances downwind by seasonal breezes. Who knows what kind of activity is taking place on exoplanets?*

A NEW KIND OF WORLD: THE "SUPER EARTH"

Astronomers thought they had a handle on the general architecture and pattern of planetary systems. The Hot Jupiters were a surprising discovery, not only in their numbers but also in their natures, but there were more exoplanetary surprises to come.

We've seen that Super Earths are one of the most common planet types among the stars, planets that fall into a category not seen in our solar system. As their name implies, these worlds weigh in at a mass between terrestrials (like Earth) and ice giants (like Neptune). Their diameters span 1.6 to more than 4 times that of Earth, and their orbits inhabit spaces near stars and far from them.

Super Earths appear to hold the majority rule in planetary systems. They outnumber Jupiter-class planets by four to one. To understand them, it helps to survey the worlds of our own solar system. The Earth is the largest rocky terrestrial, but not by much. Venus is essentially our twin in size, and it demonstrates how much atmosphere a terrestrial world can hold. With an air pressure 90 times that of Earth's at sea level, Venus simmers at temperatures hot enough to melt lead. Mercury and Mars are considerably smaller. Past the edge of the terrestrial realm reside the gas giants, Jupiter and Saturn. Their hydrogen/helium makeup has the added feature of substantial ammonia and these goliaths are large enough that their core pressures turn hydrogen into a liquid metal. But the next worlds out, Uranus and Neptune, cannot compress their hydrogen/helium gases as much, so their cores are strange oceans made up of frozen gases and water. The Super Earths represent a transition between the terrestrials and the giant worlds. Where is the tipping point between a terran and an ice giant? We do not know yet, but the answer lies within the Super Earths.

Super Earths come in a range of sizes and masses, and both these traits combine to determine their natures. The smaller ones are likely rocky, oversized versions of Earth and Venus, perhaps with thick atmospheres, depending on how close to their star they are. Some may be covered in global oceans. But the larger ones are probably of a different nature, as their water will be under great pressures, existing as superheated vapor or ice cores blanketed by deep, steam atmospheres. NASA's Eric Lopez explains, "A lot of those planets are too big to be pure rock. In the case of a world with a global ocean, that's a rock planet. That liquid ocean is a rounding error that you can't even

Right: *No one has seen a Super Earth. We can only guess at their true nature, but researchers know they measure somewhere between a terrestrial planet like Earth and an ice giant like Uranus or Neptune. This view envisions a Super Earth in the habitable zone of its star, seen from a nearby moon. The planet shares characteristics with both terrestrial and gaseous worlds. Sulfur has been detected in the atmospheres of some and this would probably tint the atmosphere yellow-green.*

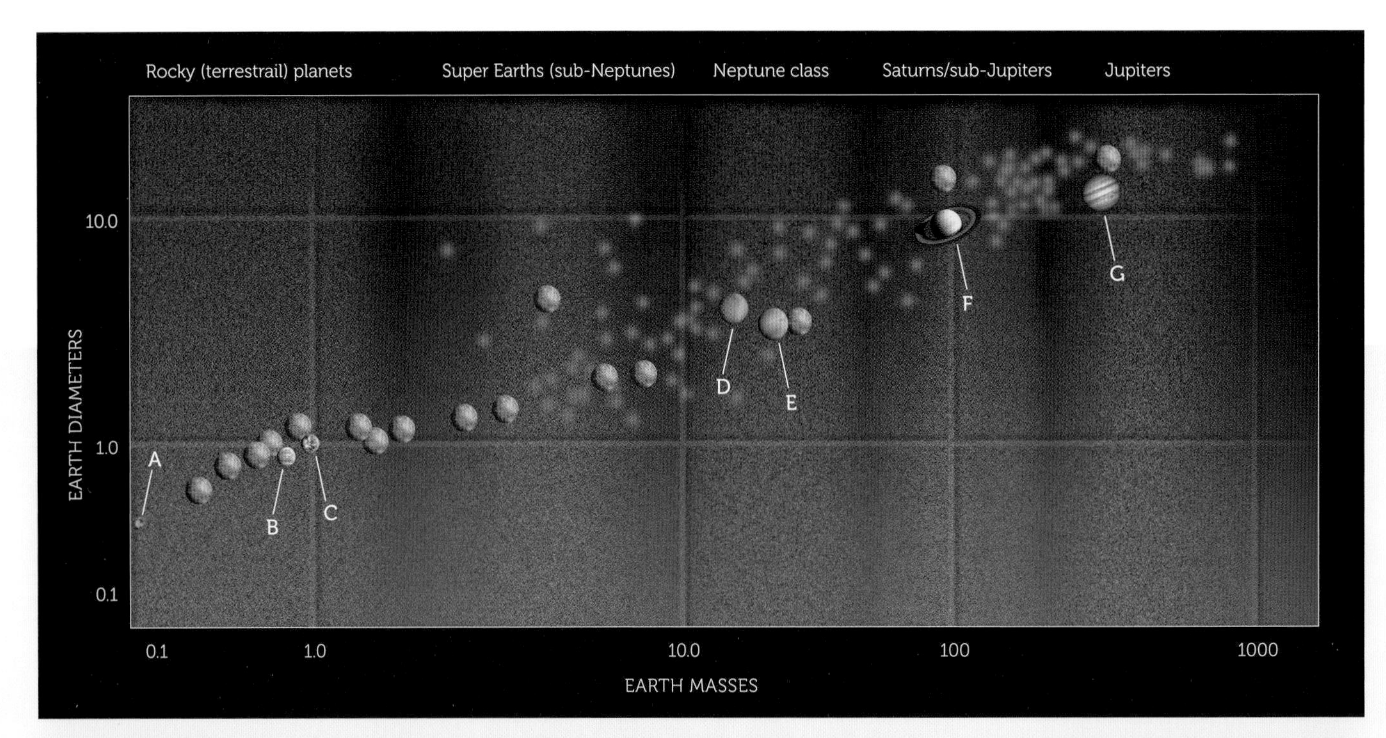
Rocky (terrestrail) planets
Super Earths (sub-Neptunes)
Neptune class
Saturns/sub-Jupiters
Jupiters
EARTH DIAMETERS
10.0
1.0
0.1
A
B
C
D
E
F
G
0.1
1.0
10.0
100
1000
EARTH MASSES

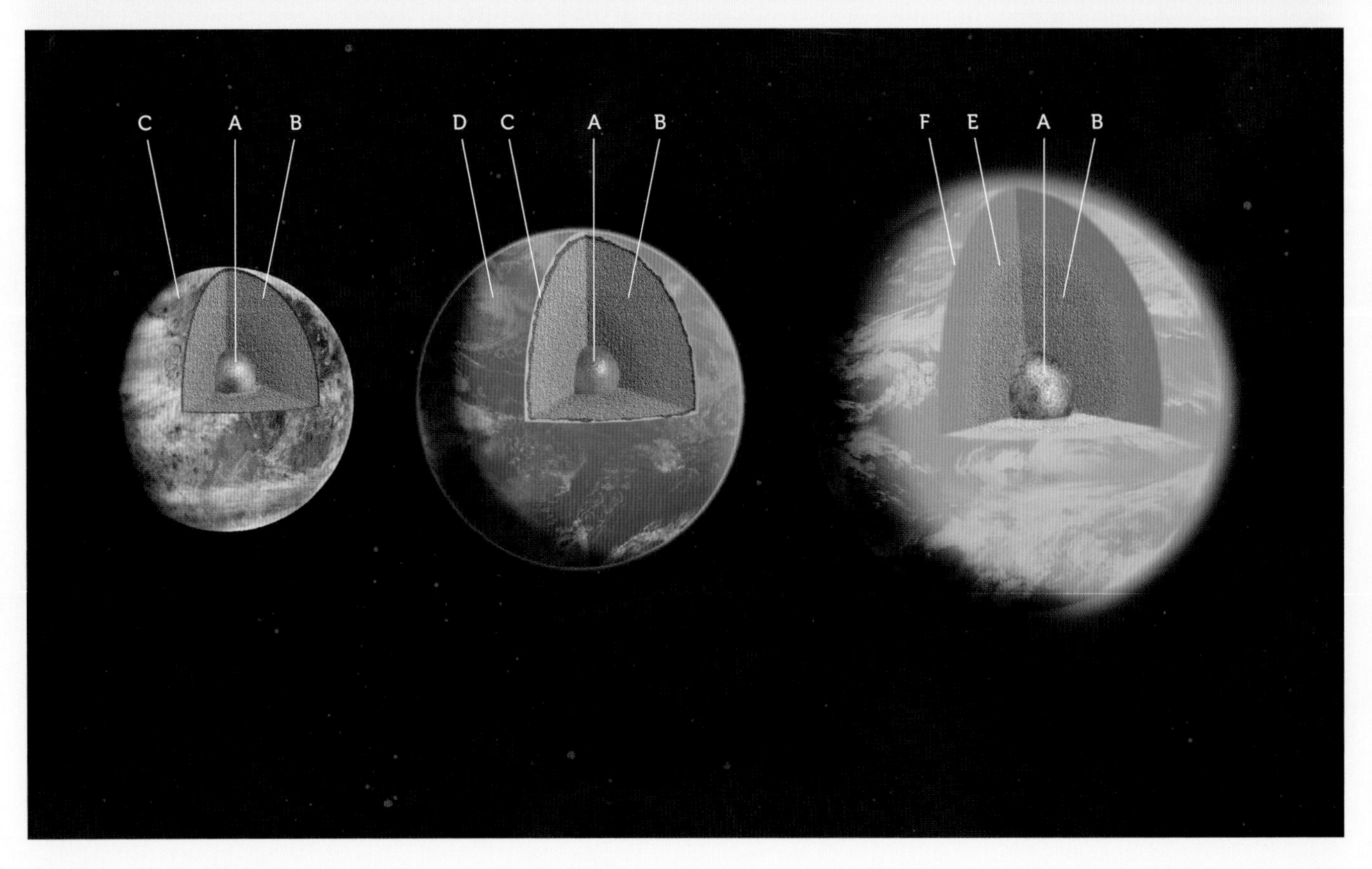
C
A
B
D
C
A
B
F
E
A
B

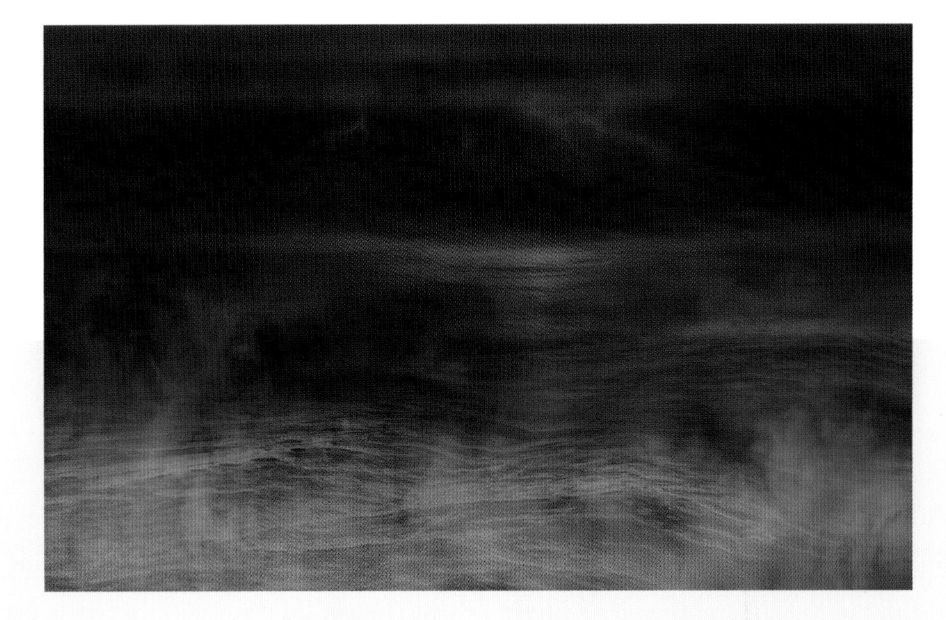

Right: *Some water worlds are Hadean steam planets, where the water vapor is nearly indistinguishable from the liquid ocean below. Farther down in the depths, the water ocean transitions to high-density ices that eventually become the consistency of rock.*

Left: *Assorted exoplanets plotted by mass versus diameter. For comparison, note the locations of planets in our solar system. Purple dots indicate some of the other discovered planets, showing trends in the population of exoplanets found so far. A=Mercury; B=Venus; C=Earth; D=Uranus; E=Neptune; F=Saturn; G=Jupiter.*

Left: *Super Earths with 1.4 Earth masses are thought to be rocky, with silicate mantles and iron cores similar to the Earth's. Larger terran exoplanets may be covered with global oceans, but may still retain Earthlike environments and interiors (center). Super Earths that approach six Earth masses may transition from rocky worlds to true water worlds, where water exists as vapor, liquid, and superdense solid ices.. A=iron core; B=rocky mantle; C=rock crust; D=water surface ocean; E=superdense solid ice; F=water liquid/vapor.*

measure. A water world is really more gaseous. The interiors of Uranus and Neptune are probably three-quarters water, so that's more what we're talking about."

NASA Goddard Space Flight Center's Avi Mandell has been studying the formation and evolution of planetary systems. His goal is to understand the factors that determine whether a system can form habitable planets. He also finds the so-called water worlds intriguing, adding, "The atmosphere can put so much pressure on that you don't even have a surface ocean; the water transitions directly from atmosphere into ice." Many of the hotter versions are "steam worlds." As water evaporates from the surface into the atmosphere, he says, "you get this incredibly steamy, cloudy thick atmosphere above hundreds of kilometers of water, above high-pressure ice where the water gets so dense and pressurized that it transitions to a solid."

A water-rich planet used to be the gold standard in the search for extraterrestrial life. But new revelations about exoplanets demonstrate that it takes more than water to provide a safe harbor for an active, living biome. The wet Super Earths do not afford any guarantees in our search for life. Vladimir Airapetian, a heliophysicist at NASA Goddard Space Flight Center, sounds a warning about the water-rich planets, even those in habitable zones. "Liquid water is not life. You need to create conditions for liquid water and conditions for protection for biology-relevant molecules that can then become more complex and eventually create life. The term 'biogenic zones' is perhaps more appropriate. Habitable zones are zones where liquid water can exist. You can have a planet with liquid water that is dead. Life is the combination of many phase transitions." Water worlds such as this are no place for Kevin Costner; they are truly alien. Although we call them "Super Earths," the larger ones may not be candidates in our search for life at all.

Of all the stars studied so far, one type has garnered the most exoplanets: the small, cool red dwarfs, or M stars. Red dwarfs are so small that their planetary systems huddle around them, making them easy targets for planet hunters using the transit technique. The red dwarf suns offer us a surprisingly wide variety of exoplanets to study.

2

KEPLER'S FINDINGS:

PLANETS, PLANETS EVERYWHERE, BUT IS THERE ANY LIFE?

The SUV-size Kepler Observatory carries 42 cameras. Those cameras combine forces to study a small patch of the Milky Way in great detail. In fact, Kepler only covers an area of the sky equal to a fist held at arm's length. But despite its focus on a tiny corner of the galaxy, by July 2015 the number of confirmed exoplanets had increased to over 1,000, with another 4,696 candidate planets awaiting confirmation. Thanks to its efforts, when we look up at the sky at night, we now know that every star we see has, on average, at least one planet circling it. With a current inventory of nearly 5,000 confirmed extrasolar planets, we might think the odds of finding an Earth 2.0 are with us. But it's hard to make an Earth. Our own planet possesses a host of variables that enable life to lodge itself here. While some astrobiologists assert that life can thrive in wildly varying conditions, there are some precursors for biology as we know it.

Left: *A large moon casts its shadow on the Hot Jupiter HAT-P-12 b. Its cool orange dwarf star is so close that the shadow is out of focus, drifting over the planet's sweltering bands of cloud. HAT-P-12 b's density is so low that the planet may bulge at the equator, much as Saturn does.*

THE EARTH'S WINNING COMBINATION

THE ORBIT

Perhaps the most important prerequisite fulfilled by our planet is that it orbits in the Sun's habitable zone (or HZ). The habitable zone is the distance from the star at which a planet can hold liquid water on its surface. It is not too far, nor too close; not too hot, nor too cold. Like the fairy tale, it is the "just right" Goldilocks zone (see "Living the good life (in the Goldilocks zone)" feature on pages 24–25). But in order to have liquid water on any surface, there must also be air pressure. Too much air, and the atmosphere will pump up temperatures to the point where water becomes steam. But at the other extreme, if the air pressure is too low, water cannot remain a liquid and turns into vapor or freezes to solid ice, like the ice moons of our own outer solar system. The majority of the Earth's surface can—and does—sustain liquid water.

THE MAGNETOSPHERE

Adding to the life-enabling list, the Earth's core is large enough to retain some of the heat from its formation, along with heat from radioactive materials. This internal heat keeps the metallic core molten, and the Earth turns quickly enough to set up currents within that molten metal. Those currents generate a cocoon of energy fields called a magnetosphere, which becomes a protective bubble around the Earth, deflecting most of the solar radiation away from the surface. (Earth's twin, Venus, probably has a molten core similar in size to our own, but the planet spins so slowly that it cannot set up currents within that liquid metal, so the planet generates no organized magnetosphere.)

THE RIGHT WATER/ROCK BLEND

Our search for Earth 2.0 ideally includes worlds with a mix of liquid water and continental masses. This mingling brings a rich diversity of life to our home world, something biologists call the "edge effect," which occurs at the interface between two very different biomes. For example, the desert regions of the Sinai Peninsula abut the waters of the Gulfs of Aqaba and Suez, as well as the Red Sea, and each of those two environments, land and sea, play host to their own balance of living things. But at the tidal zone between, the diversity of life skyrockets. Such may be the case—indeed may be a life requirement—on exoplanets as well: the interplay of land and water may play a critical part in biodiversity and advanced life. Still, there are water worlds out there, and their seafloors may be the home of underwater gardens of rich life, oases whose inhabitants may have made their way to the surface.

PLANETARY COMPANIONS

Judging by the exoplanet arrangements we have charted so far, the architecture of our solar system seems rare. We have no Hot Jupiters close to the Sun to disrupt the orbits of terrestrials in the habitable zone, but we do have a massive gas

OUR OWN PLANET POSSESSES A HOST OF VARIABLES THAT ENABLE LIFE TO THRIVE HERE.

Above: *The Earth's magnetosphere creates a protective force field, funneling solar radiation toward the poles and away from the planet. This magnetic bubble issues from our molten iron core. Currents within it, driven by the Earth's spin, set up magnetic fields. Venus likely has a similar core, but its slow spin prevents it from generating such a protective force field.*

giant much farther out. Jupiter orbits in a gravitational sweet spot: its influence works to clear the inner solar system of large asteroids. The Sun's powerful gravity tugs at asteroids and comets, tending to pull them inward toward the region of the terrestrial planets, but Jupiter's orbit is in just the right location to set up resonances with incoming asteroids and comets, essentially creating a shield for the inner planets. Jupiter's interaction with asteroids and comets preferentially tosses them toward the outer solar system rather than allowing them to fall Sunward, where their impacts on Earth might provide reruns of the event that wiped out most of the dinosaurs. No such planetary arrangement has yet been found in other solar systems.

The Earth has yet another important planetary companion: its own Moon. The Moon is so large compared to Earth that the two practically constitute a double planet (as Pluto and its largest moon Charon do). The Moon plays a critical role in keeping the Earth's axis stable. The Earth's spin is tilted by about 24°. This tilt gives us our seasons and helps to govern our global weather patterns. Mars has a similar axial tilt, but within 50,000 years, Mars will be tilted on its side. Like a spinning top on a counter, Mars will rock back and forth. This rocking motion is called precession. The Earth's Moon

Right: *Jupiter looms above a volcanic caldera on its moon Io. The gas giant is the largest planet in our solar system.*

Above: *The early Moon was a world torn asunder by molten rock and pummeled by meteors. The Earth's large companion keeps our world from wobbling in its spin.*

is heavy enough that it stabilizes the precession in our spin, so that changes in climate and seasons are gradual. The Moon also sets up tides, which may have played a key part in the establishment of life. Finally, early in the Earth's formation, the Moon may have stripped away some greenhouse gases, keeping our air thin. If it had not, our fate may have been similar to pizza-oven Venus.

A PLANETARY RECYCLER

The Earth has several recycling systems that are built into its life cycles. The most obvious is the water cycle. Clouds condense; rain falls into rivers, lakes, and seas; and the water evaporates to begin the cycle again. Similar cycles are found on other planets and moons involving other volatiles like ammonia (in the gas giants) and methane (on Titan). The Earth's water cycle naturally purifies its water, keeping it in balance for biological processes.

But the Earth has other recycling functions, and these appear to be rarer among the worlds we've explored. The most remarkable is the process of plate

Above: *The Earth's crust is fractured into sections called plates. Some plates abut each other, lifting mountain chains. Others slide under each other, melting the rock and freeing up trapped atmosphere (often through volcanoes). Do such plates exist on terran exoplanets?*

tectonics. The Earth's crust is fractured into sections of movable provinces that float on the planet's mantle. As they move, some crash into each other, raising mountain chains (the most famous of these is the Himalayas, which rose as India moved northward, ramming into Asia). In other places, plates subduct, or slide under each other.

Plate tectonics sculpt our environment in two important ways. First, they act as an atmospheric recycler. The rocks on a terran planet behave like a chemical sponge, soaking up elements of

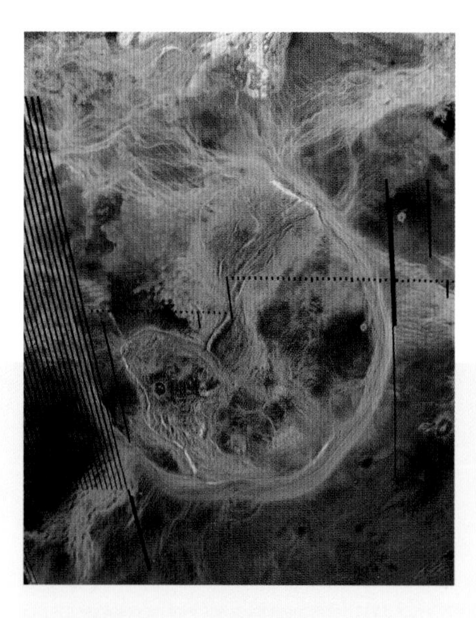

Jumbled land forms compressed by sinking mantle

Coronae on domed crust over rising mantle

Volcanoes over upwelling mantle sites

atmosphere and locking them into the surface. But as plates subduct, rock melts and the gases are freed, often venting through volcanoes and back into the atmosphere. This recharging was present on ancient Mars, which hosts the largest volcanoes in the solar system. But when those volcanoes shut down (Mars has no plate tectonics today) the atmosphere drifted away, eroded by the Sun's radiation and the planet's low gravity.

The second positive aspect of plate tectonics has to do with the minerals in the soil. Early naturalists believed the Earth was only a few thousand years old, and one of their primary arguments was that over time, surface minerals would erode and wash into the ocean basins. Plate tectonics were unknown at the time, but the process is responsible for recycling minerals as well as air, recharging the environment over eons.

Not all terran exoplanets may have processes like plate tectonics. Some models suggest that rocky planets larger than Earth have crusts too massive to fracture into plates, so the recycling of minerals in the rock and the gases in the air may not happen at all, or may happen only at volcanic sources. NASA geologist Eric Lopez says that there are two schools of thought on whether an exoplanet—especially one larger than Earth—has plate tectonics. "It really depends on what is the viscosity [thickness] of the molten rock, or magma, beneath the surface. Low viscosity gives you a stagnant 'lid' of rock, like Mars has today. The planet may have a couple of hot spots, but a frozen-in-place crust. But if the magma is less viscous, you might get tons of tectonics with thousands of microcontinents." The problem facing researchers is that they have no idea what kind of viscosity is typical on Super Earths. Those who model planetary structures have come up with two competing theoretical models, says Lopez. "One predicts that volcanism should go up by a factor of a thousand, while the other predicts that it will turn over and go down by a factor of a thousand. There are no lab experiments to show which of those is right."

To see how rare these tectonic forces may be, we need to look no farther than our own solar system. Bracketing the Earth at the edge of the Sun's habitable zone are Venus and Mars. For its part, diminutive Mars is too small to have held on to a substantial atmosphere, but it seems to have had the added problem of a lack of tectonics. Studies of magnetic

Above: *Alien tectonics: Venus exhibits strange circular folds and faults, called coronae, that may indicate a type of tectonics. Other areas are jumbled and raised, while still others host possibly active volcanoes. The diagram at right shows proposed models of the forces at work.*

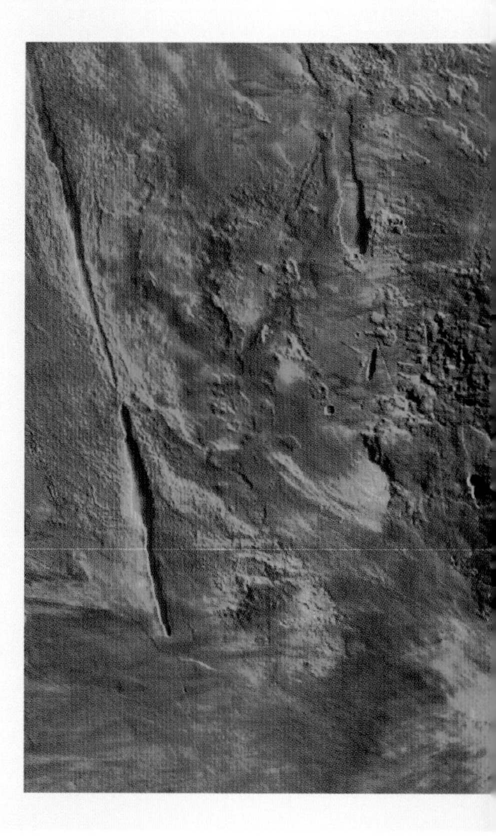

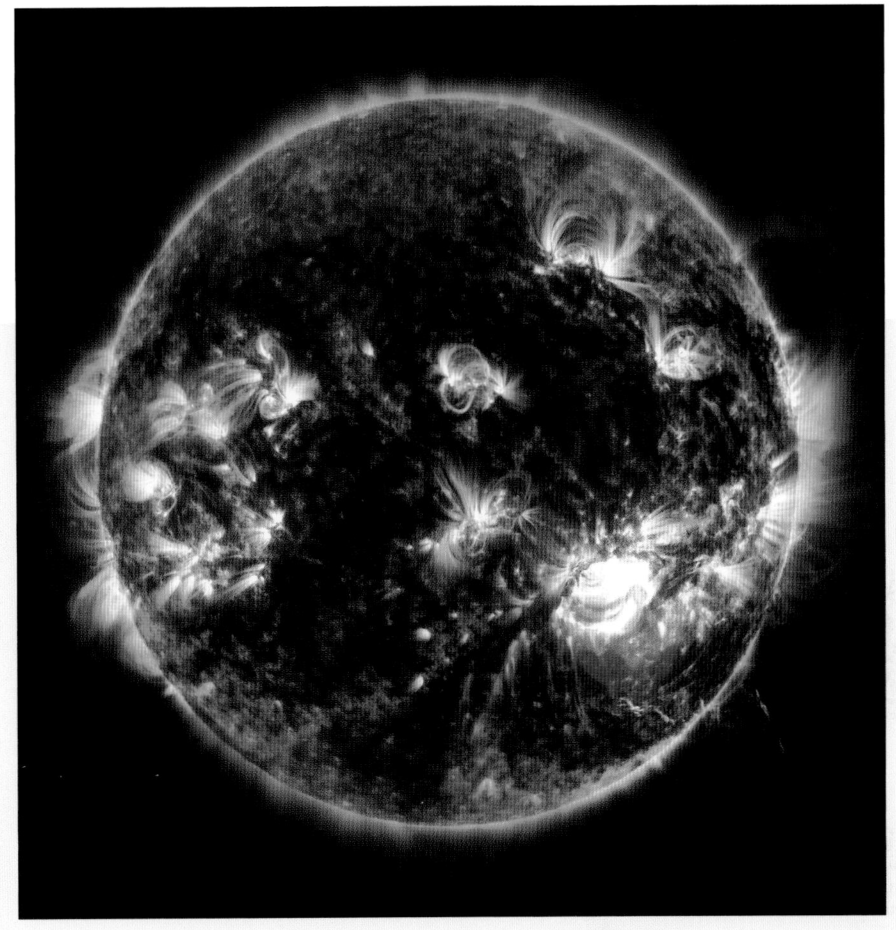

Right: *This image of our own star, the Sun, was taken by the Solar Dynamics Observatory in ultraviolet light. The Sun is a G-type star, a stable star on the main sequence.*

Below: *Mars may have had plate tectonics that shut down early, leaving clear conduits for volcanoes to grow into the largest mountains in the solar system. Eventually, that volcanism died out. Seen here is the magnificent Arsia Mons.*

patterns in the rocks imply that tectonics were at work early in the red planet's history, but any drifting crust must have become silent early in the Martian past. No tectonics sustained volcanoes or recirculated the atmosphere. At the cold outer edge of our habitable zone, Mars had no chance of retaining the kind of active surface biome that Earth possesses.

Venus has tectonics of a different kind, perhaps columnar in nature. Its volcanoes may well be active today, but any water has boiled away. The Venusian crust seems to have upwelled in great arenas and oval regions, with material rising in the center and diving back down into the crust on the outer rim. This form of tectonics may not produce the revitalizing force of Earth's tectonics.

THE TYPE OF STAR

The type of star in a planet's sky may also be a critical factor. As we saw in Chapter 1, the giant and supergiant stars burn themselves out in a few billion—or even a few million—years. It is widely thought that these stars are not steady enough to support stable conditions long enough for life to arise. On the other end of the scale, the most common star class in the galaxy is the M star, or red dwarf. These ancient, cool stars, often roughly the size of Jupiter, possess entire families of planets, but red dwarf suns are not as stable as our own. Although they are long-lived, these cool suns have a nasty tendency to belch out deadly solar flares. Planets of such cool stars must orbit closely to be in the warm habitable zone, but their tight

Left: *The galaxy NGC 6217 is a barred spiral, similar in shape to the Milky Way in which we live.*

orbits place them in the danger zone for these flares.

Our Sun is comparatively tame. Its output is stable and fairly predictable on an 11-year cycle, and it is hot enough that our habitable zone is fairly distant from its flares and prominences. Some biologists suggest that we have just the right kind of star for the gradual unfolding of life on a global scale.

TYPE OF GALAXY

The universe is organized like a vast web or sponge, with many hollows and empty chambers. Its galaxies are arranged like piles of soap bubbles, with the galaxies themselves clinging to the surfaces of the "bubbles." The bubbles surround empty voids and are attached to each other via filaments of stars and galaxies.

The stars of some galaxies are loosely packed in configuration, drifting in an amorphous cloudlike flotilla. Other galaxies are more organized, great whorls of suns in vast disks, many with spiral arms extending out hundreds of light-years.

The densest of galaxies emit deadly gamma-ray bursts (GRBs), mysterious blasts that may come from matter falling into black holes or may mark the merger of a black hole and neutron star. GRBs can last from milliseconds to hours, but in that short time, these high-powered explosions pour out hundreds of times the energy of a supernova, enough to destroy all life in entire solar systems through vast regions.

One such event occurred on July 9, 2005. Despite the fact that it took place a billion light-years away, a wave of faint gamma rays passed through our solar system. The High Energy Transient Explorer satellite detected the event, which is called a "short duration GRB." The fading wave of energy lasted just 70 milliseconds, but astronomers across the world were able to combine forces and tease out some of the details. The gamma rays marked a violent event:

SOME BIOLOGISTS SUGGEST THAT WE HAVE JUST THE RIGHT KIND OF STAR FOR THE GRADUAL UNFOLDING OF LIFE ON A GLOBAL SCALE.

Above: *A neutron star begins to fall apart as it spirals in toward the accretion disk of a black hole. When it finally succumbs to the black hole's mighty gravity and merges with it, the dying star will unleash a titanic explosion of energy, seen from a distance as a Gamma Ray Burst.*

a 25km (16 mile)-wide corpse of a collapsed sun, a neutron star, self-destructed as it was swallowed by a black hole. Though the event was subtle at our distance, any nearby solar system would have been sterilized by the cataclysmic explosion.

In corners of the universe where galaxies are densely packed, GRBs from one galaxy can actually destroy life in adjacent ones. Some researchers suggest that galaxies inhabiting the quiet voids are more likely to host Earthlike worlds with life. Our Milky Way galaxy, a great pinwheel of 100–400 billion stars, resides in just such a location.

The great spiral of stars making up our galaxy spans some 120,000 light-years across. Within this structure, interstellar gas and dust drift among the star-birthing nebulae. Young and middle-aged stars glow throughout, but the majority of stars in the central bulge are older. A supermassive black hole at galactic center attracts a cyclone of gases and stars around it. The object pumps out prodigious amounts of X-rays and other radiation.

OUR PLACE IN THE MILKY WAY

The Milky Way is a barred spiral galaxy (a spiral shape with a heavy "crossbar" of stars through the center). Over three-fourths of all galaxies are spiral. More than half of these spirals scatter their suns in multiple arms. Unlike many galaxies, the Milky Way has avoided many collisions with galaxies in the past 10 billion years. This is significant: when galaxies drift into each other, the collision causes a wave of deadly supernovae, exploding stars that emanate deadly radiation.

Black holes may be quite common at the centers of galaxies. The black hole at the center of the Milky Way is only moderately active, providing just enough energy to keep the center of the galaxy active, but not so energetic as to destroy massive amounts of stars through gravitational pull or eruptions of gamma-ray radiation.

Stars are packed closely together toward the center of the Milky Way and

THE EARTH AND ITS PLANETARY SIBLINGS ARE WELL PLACED IN A QUIET, RESOURCE-RICH NICHE OF A VAST AND COMPLEX GALAXY.

Left: *Our entire galaxy may have a constrained habitable zone (in green) populated by stars able to support biomes within their planetary family.*

our Sun orbits in a less dense region where deadly radiation from other stars does not impinge upon it. With the dangers of our galactic hub, some researchers estimate that stars more than 6,500 light-years from the center are more likely to harbor life. The Earth orbits 28,000 light-years from the center, far away enough to elude the lethal gamma radiation coming from the galaxy's heart. Additionally, the older stars in the center of the galaxy are starved of the dense metals and minerals that serve as building blocks of terrestrial planets. Our planetary system orbits the galaxy far enough from the center to have access to the heavy elements generated in the explosions of supernovae.

A whopping 95 percent of Milky Way suns may not be able to sustain habitable planets. Many worlds circle the galaxy in paths that take them through the deadly spiral arms of our starry pinwheel. Any star that passes through one of these glowing arms is subject to deadly radiation from closely packed stars. Our own solar system orbits in a fairly circular path and it is also in sync with the rotation of the rest of the galaxy, keeping it in a quieter space between the spiral arms. The Earth and its planetary siblings are well placed in a quiet, resource-rich niche of a vast and complex galaxy.

Our list of biological prerequisites is limited by the fact that Earth's life is the only example that we know. The interrelated plants, microbes, and larger fauna all depend on the environment

A SHORTLIST FOR EARTH'S ENVIRONMENT

While the list below may not represent absolute prerequisites for life, it provides a survey of the factors that contribute to the benign, biologically friendly environment of Earth. Life may be present on worlds lacking some or most of these characteristics.

- Just right star
- "In the zone" (Goldilocks distance from the Sun=liquid water)
- Jovian asteroid deflector
- Gravity + temperature
- Just right moon
- Plate tectonics and volcanoes that recycle minerals and air
- Radiation shield (magnetosphere)
- Galaxy type
- Place in galaxy

The complex mix of variables that enable life to thrive on Earth provide a list for astrobiologists—scientists who study the possibilities of life in the cosmos. This list forms a basis for evaluating just how Earth-like a planet is. The list is called the Earth Similarity Index.

> SEARCHING FOR PLANETS LIKE EARTH COMES DOWN TO A NUMBERS GAME . . . ASTRONOMERS HAVE BEEN ABLE TO COME UP WITH AN ESTIMATE . . .

from which they arose. Could exoplanets count among themselves a similarly favored world? Our narrow view of life's necessities was broadened with the advent of space exploration and, in particular, the discovery of ocean worlds within our own solar system. A heavenly host of icy moons contain subsurface oceans and many are subjected to the pull of gravity from nearby moons and their planet (tidal heating). This combination opens up possibilities for life within our own system of planets and moons, and it broadens the territories we may consider life-friendly within the myriad of exoplanets beyond our own solar family.

The Earth Similarity Index (ESI) gauges the similarity of a planet compared to Earth. It lists several factors, including the planet's diameter, density, gravity, and its range of surface temperatures. The ESI spans from zero to one, with the Earth being one. In our own solar system, we have three planets roughly in the habitable zone: Venus, Earth, and Mars. The ESI for Venus is 0.444, while the ESI for Mars comes in at 0.697.

Searching for planets like Earth comes down to a numbers game. Where are they located? What is their ESI? How many are there? With growing confidence, astronomers have been able to come up with an estimate for the last crucial question: the Milky Way galaxy may host as many as 11 billion planets the size of the Earth. That astounding number must be kept in perspective, however: not all are in the habitable zone and many orbit stars that likely destroyed their atmospheres early on. Still, it's a big number, and it inspires us to continue the search throughout our own solar system and beyond.

BREAKING THE STEREOTYPES

What will the "biological prerequisite lists" of various exoplanets look like? Will very different conditions still lead to living systems? After all, life on Earth began in an environment quite different from the one we enjoy today. Opaque clouds blanketed the sky for millennia, and that

Below: *The Earth's volcanism recycles minerals into the ground and gases into the atmosphere.*

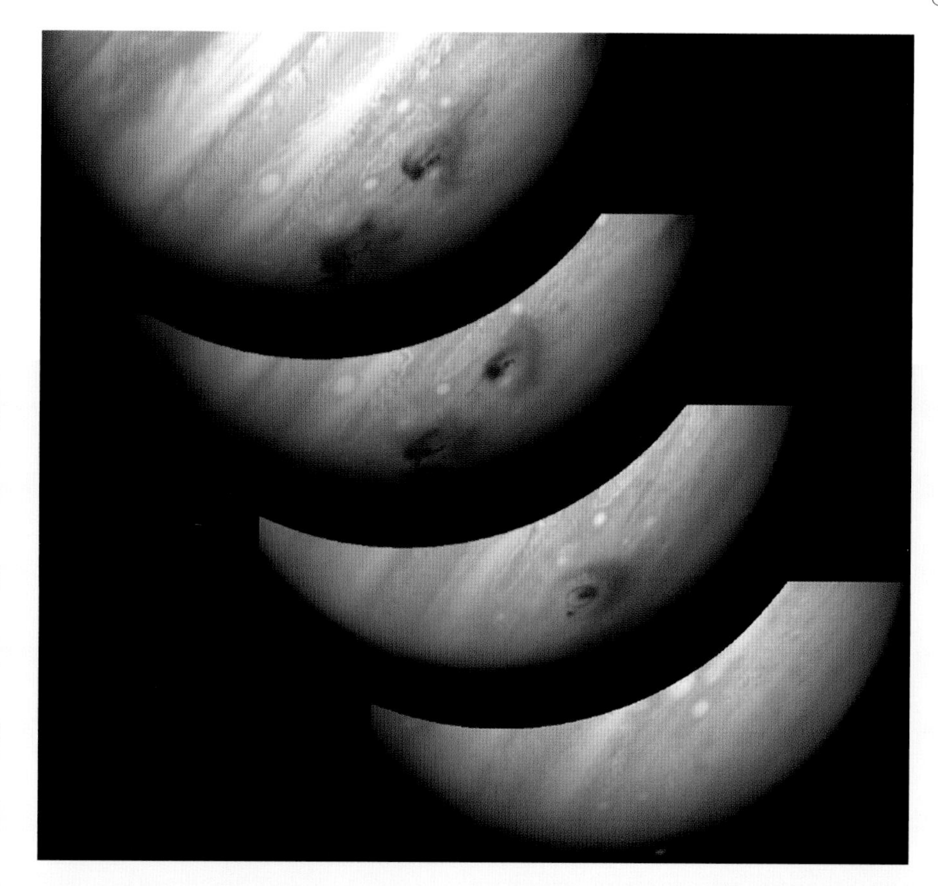

Above: *Jupiter's role as asteroid and comet deflector was dramatically shown when Comet Shoemaker-Levy 9 broke apart and pummeled the planet, leaving a trail of dark blemishes in the seared atmosphere. This remarkable sequence was taken by the Hubble Space Telescope.*

sky had no oxygen, but rather a mix of elements that seem like poison today. Within this setting, the first creatures arose, life that would have died in the presence of the oxygen that most Earth life now cherishes. Anaerobic bacteria populated the seas, eventually giving rise to more complex life such as colonies of microbes that formed mats on the seafloor. Life emerged using photosynthesis, as plants do today, to breathe and make energy. Their waste product was oxygen and as the oxygen piled up, life-forms learned to make use of it. The early Earth was a truly alien world. Continents of lava remained black, as no oxygen was present to "rust" the minerals in the rocks. The seas would have been a deep green due to non-oxidized iron within the water.

Life on Earth relies on carbon and water—humans are carbon-based life-forms—but do other environments offer construction materials for life? A world in our own solar system is providing us with a fine laboratory for such important and fascinating questions. That world is Saturn's planet-size moon, Titan.

Titan seems to be a hostile place for life. Its nitrogen-methane atmosphere is the second-densest atmosphere among all the solid bodies of the solar system (second only to Venus) and temperatures on the surface hover at −178°C (−288°F). While that seems cold, Titan is much warmer than Saturn's nearby ice moons, whose daytime temperatures dive 25°C (45°F) below that.

Titan's clouds rain cryogenic methane, a bizarre precipitation that carves river valleys which drain into liquid-filled basins, some as large as the Black Sea. But methane has some advantages over water for life, says astrobiologist Christopher McKay of NASA's Ames Research Center. "Methane is chemically benign compared to water, which tends to break biomolecules apart. Also, Titan does not suffer from ultraviolet radiation, something which terrestrial life must constantly combat. Finally, in Titan's low temperatures, life will decompose slowly."

A second kind of precipitation may have even more bearing on Titan life: a steady drizzle of hydrocarbons. This organic "soot" combines with methane to form complex compounds, the raw material of life.

Methane and hydrocarbons aren't the only things falling from Titan's cloudy

skies. Sunlight and radiation from Saturn's magnetic fields break up nitrogen and methane molecules. When these fragments recombine, they create a compound called vinyl cyanide. This is important in the search for life because it tends to accumulate into films like those found in the protective membranes of terrestrial living cells.

This news has exciting implications for the possibiities of life on Titan, however remote. Over time, an astounding 10 billion tons of vinyl cyanide may have accrued in Titan's largest methane seas, Ligeia Mare and Kraken Mare. And while a cell membrane is certainly not a guarantee of life, it is likely one of the prerequisites.

As Titan's skies produce this smorgasbord of organic materials, they rain down into the methane lakes, congregating into more complex combinations, perhaps leading to the precursors of life. And if Titan's severe environment could conceivably support life, other exoplanet landscapes may as

Above: *The bizarre methane and hydrocarbon rainfall on Saturn's moon Titan may provide life-friendly chemistry.*

Above: *Technicians load the Kepler Observatory onto its upper stage prior to launch. Kepler uses the transit technique, staring at the rise and fall of starlight to detect planets passing in front of stars.*

well. The radiation and heat from stars bathe some planets in the kind of energy necessary for life. If the planet is far enough away to dodge solar flares and close enough to promote biochemical reactions, then life may arise in many places.

With advanced orbiting and Earth-based observatories coming on line, astronomers are discovering a rich variety of new solar system arrangements, new planets, and new kinds of planets. We have seen the Hot Jupiters from the early days of exoplanet exploration. They were the "easier" discoveries, the planets near enough—and massive enough—to make themselves known dramatically by the way they move their suns.

New techniques have brought more nuanced results to the search. Observers have glimpsed Jupiter-size planets orbiting in habitable zones. Added to their number are smaller, rocky terran planets, and in between are the mysterious Super Earths, hybrids of terrestrials and great ice giants.

THE KEPLER ERA

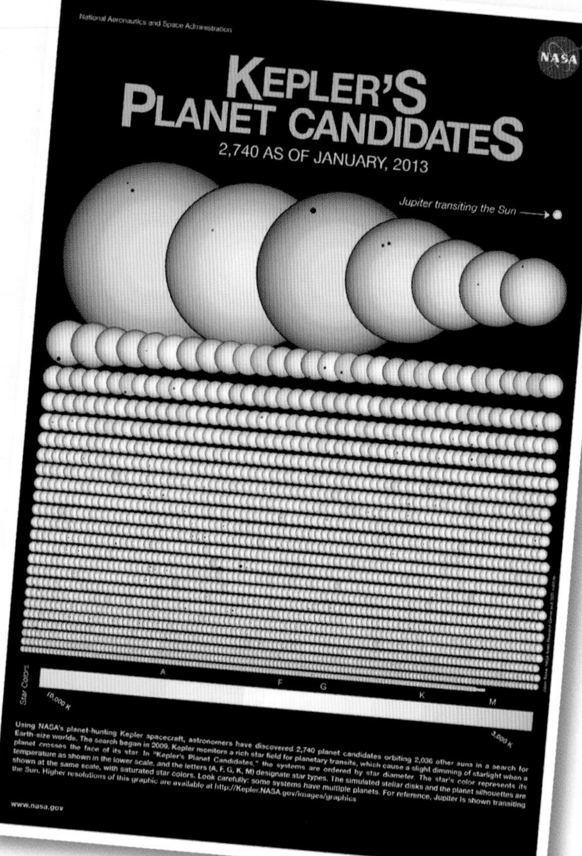

Above: *Some planetary systems discovered by Kepler's transit technique. Stars are all to scale, as are the transiting planets in front of each, and arranged by diameter. Color indicates temperature (see scale below). The Sun is shown at upper right, with Jupiter transiting.*

The Kepler Space Telescope, tasked with studying stars within a small swath of our own galaxy, has revealed thousands of worlds, both familiar and alien. Early in the mission, Kepler picked the low-hanging fruit, the large planets easiest to detect as they passed in front of their star. But as data accrued, the Kepler team was able to tease out more subtleties in the data. What they found was astounding: bizarre planets where rock turned to vapor; Jupiters orbiting their suns as closely as Mercury orbits ours; weird Super Earth water worlds unlike anything in our solar system, where global seas turn to vapor under great pressures.

Kepler's transit technique may represent the most dramatic advancement in the hunt for exoplanets so far. As of the fall of 2019, some 4,774 planets are either candidates (still to be confirmed by further study) or confirmed. Padi Boyd, head of NASA's Exoplanet and Stellar Astrophysics Laboratory, says, "Kepler's goals were straightforward: It was tasked with answering how common is a planet like Earth around a star like the Sun? It turns out that planets are everywhere."

Thanks to Kepler and other observatories, we now know that not only are planets abundant throughout the galaxy, but they are also diverse. The stars hosting planetary systems are also diverse, and the architectures of those systems vary in composition and arrangement. Dozens appear to be rocky terrestrial worlds orbiting within the habitable zones of their stars.

Despite its stunning successes, the transit method has several disadvantages. Planetary systems tend to orbit around their star in a flat disk. The transit method only works with solar systems whose planetary disk crosses the face of the star when seen from Earth's perspective. Only 10 percent of exoplanets in close orbits pass directly through the line-of-sight to their star as seen from Earth. The farther away the planet orbits from its sun, the less likely it is to cross its face. In the case of a world circling the same distance as the Earth does from the Sun (one AU), the probability of a sun-crossing transit is only 0.5 percent. This means that for every 200 exoplanets out there, only one transits its star. All the rest remain hidden to the transit technique, orbiting in directions that keep them from crossing the face of their sun (from our viewpoint). For example, a star whose pole points directly at the Earth will have a system of planets arranged like a bull's-eye, with the planets never crossing in front of the star's lit face.

Another limitation of the transit method is that not all planets orbit in a

KEPLER'S TRANSIT TECHNIQUE MAY REPRESENT THE MOST DRAMATIC ADVANCEMENT IN THE HUNT FOR EXOPLANETS SO FAR.

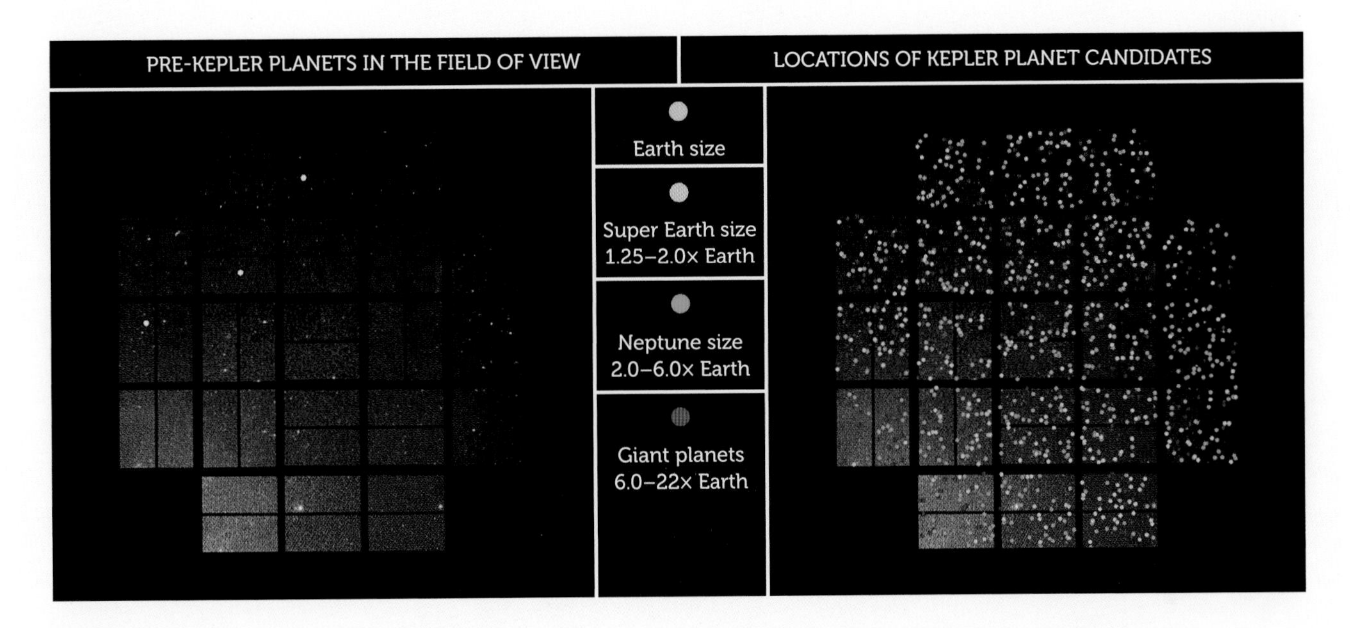

BILL BORUCKI, INVESTIGATOR OF THE KEPLER MISSION

Just 30 years ago, many astronomers believed exoplanets should be relegated to the realm of science fiction—but one determined NASA Ames researcher thought otherwise. His name is William Borucki.

"In the eighties," Borucki remembers, "it was pretty clear that if you want to find planets around other stars, you do it with astrometry. At the time that was the idea: to watch for the movement of a star. But some of my colleagues suggested that I try something else." Borucki read a paper on photometry, the study of light levels. Although the paper's authors had made some mistakes, their work presented a logical explanation of how to use photometry to detect distant planets. "So it was just a matter of building a photometer a thousand times better than anyone had ever built."

Borucki set out to fabricate a new generation of instrument to search for exoplanets. After many setbacks, disappointments, and false starts, NASA okayed the Kepler Observatory, a spacecraft that has discovered more planets than all other techniques combined. "We thought, for example, that the orbits would have to be circular or they'd be running into all kinds of things. But they're not. And would you dare have Jupiters in a four-day orbit? Impossible! So all we've seen is surprise after surprise after surprise." Kepler's thousands of exoplanetary revelations are all thanks to one tenacious man who wanted to gaze at starlight.

Above: *Left: Before the Kepler Mission, a handful of exoplanets were known in this view of the sky. Right: A plot of Kepler planetary candidates as of 2011, just two years after mission start. The spacecraft continued operation until November 2018 and the list of confirmed and candidate exoplanets grew substantially.*

Below: *Bill Borucki (pointing) is the genius behind the Kepler Space Telescope's transit technique.*

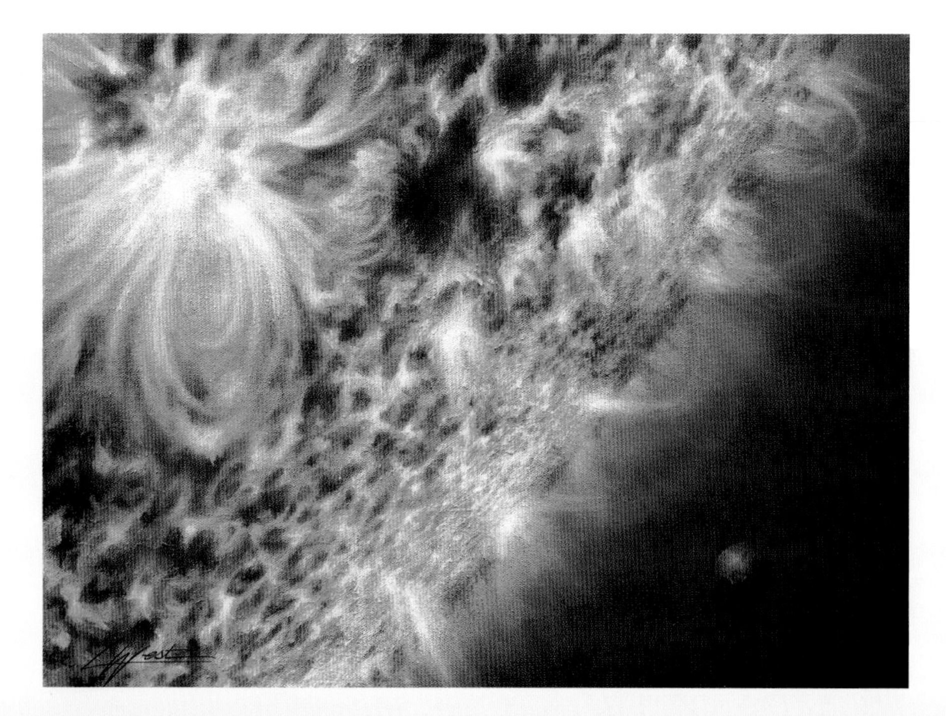

Left: *A planet transits the face of its star. Such crossings can tell astronomers much about a distant planetary system.*

Below: *Despite their small size, M-type stars, called red dwarfs, are notorious for belching out deadly flares.*

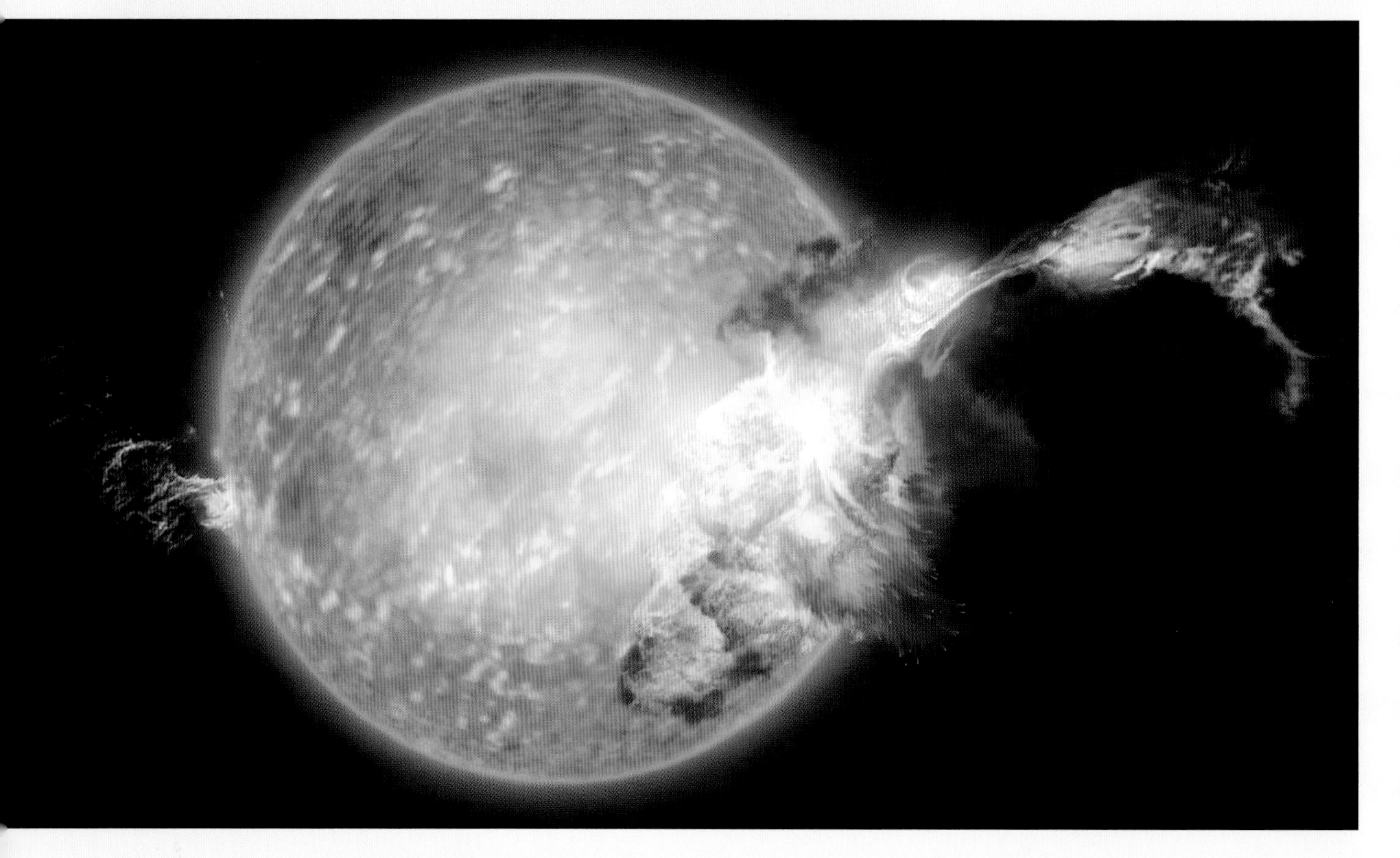

Above: *The Hubble Space Telescope is released by the Space Shuttle after a repair and refurbishment mission in May 2009 (STS 125).*

perfect plane; some are slightly offset. In the case of our own solar system, an astronomer in the nearby Alpha Centauri system might see only two or three of our terrestrials at once: Mercury, Venus, Earth, and Mars do not cross the Sun in quite the same alignment. The Earth's orbit is tilted (inclined) by about 1.5° and the Venusian inclination is close to 4° off the plane of the rest of the planets.

Yet another problem arises with stars that vary in brightness (or luminosity). The light from red giant stars like Betelgeuse ebbs and flows, confusing dips in light that might indicate the transit of a planet. Similarly, red dwarf stars are so variable in their brightness that they have earned the name "flare stars." These Jupiter-size suns often belch forth solar flares, increasing their brightness and wreaking havoc with the subtle light level readings used in the transit technique. It takes many observations to tease out the regular dimming of a transiting planet, but with patience it can be done.

OBSERVATORIES OLD, NEW, AND YET TO COME

Knowing planets are out there is one thing; finding them is yet another. Ground-based observatories are challenged by hosts of barriers to their view, including the Earth's atmosphere and light pollution. As the new millennium opened, it was time to take to the skies.

At the close of 2006, the French space agency CNES (Centre national d'études spatiales) combined forces with ESA to loft the CoRoT (Convection, Rotation et

Above: *This artist's impression shows the planet K2-18b with its host star and an accompanying planet in this system. K2-18b is now the only Super Earth exoplanet known to host both water and temperatures that could support life (see Chapter 4).*

Transits planétaires) space telescope. It was the first mission designed to find transiting planets and it did not take long: CoRoT achieved its first two exoplanet discoveries in its first year of operation (2007), the two Hot Jupiters Corot-1 b and 2 b. Among its other discoveries were the first rocky exoplanet, Hot Jupiters in highly eccentric orbits, and several multiple-planet systems. CoRoT was retired in 2013.

Several observatories already in operation have joined Kepler in studying exoplanets. The Hubble Space Telescope, operated by NASA, ESA, and the Space Telescope Science Institute, has become a powerful tool in the search for exoplanets and life that may inhabit them. Along with the Spitzer Space Telescope, Hubble has identified water vapor in the atmosphere of many Super Earths. Hundreds of known exoplanets fall into this mass range. Astronomers are struggling to characterize Super Earths and evaluate the chances of their supporting environments with active biology.

The spacecraft also charted a system of three planets orbiting in their star's habitable zone that seemed to lack the kind of extended atmosphere seen on hydrogen/helium-rich giants like Neptune. These Super Earths, although large, may well be rocky, with atmospheres containing the kinds of gases found on terran planets. Hubble's sky-high vantage point above Earth's atmosphere gives it clear enough vision to sense the subtle changes in starlight that indicate the presence of

GETTING THE SPECTRUM FROM A BLOB OF LIGHT

In most cases, a star appears as a point of light to a telescope, and any planet is invisible. But there is still a way to "see" a planet's atmosphere. Observers chart the light of the star before a planet transits. During transit, a second spectrum is taken. The two are compared and the difference reveals constituents of the atmosphere as the starlight passed through it. Observers on Earth can sample the spectrum of a star before a planet's transit. As the planet crosses the star's face, it blocks some of the starlight with its solid surface, but its atmosphere blocks light partially. When the star's spectrum is subtracted from the transit light, astronomers can tell what constituents make up the atmosphere of the planet.

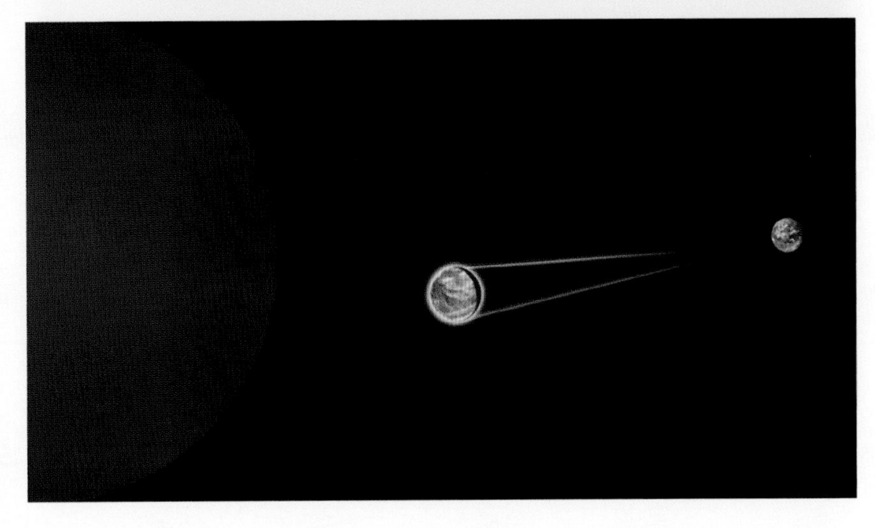

Right: *Both the Hubble and Spitzer have contributed to studies of exoplanet atmospheres by subtracting the light of a transit from the starlight before the transit.*

Above: *The Spitzer Observatory saw the universe in a different part of the spectrum than Hubble did. Its infrared imagers could detect heat, revealing previously unseen details.*

an atmosphere. In fact, Hubble was the first observatory to image an exoplanet in visible light. The planet orbits the star Fomalhaut, a sun surrounded by a disk of gas and dust. The discus-shaped cloud had a fuzzy outer edge, but something was keeping the inner edge sharp; at that inner edge of the disk, Hubble revealed a point of light a billion times dimmer than Fomalhaut circling the star. Hubble has also shown astronomers planets forming around new solar systems and other new exoplanets embedded within disks of debris. These planets caught in the act of their birth are called proplyds. Several examples of other exoplanets have been found at stars such as TW Hydrae and Beta Pictoris.

In Hubble's wake, the Spitzer Observatory rose to the occasion in fine style by charting the heavens in the infrared part of the spectrum. In fact, Spitzer was the first telescope to detect the light of exoplanet atmospheres directly. Spitzer is revealing in that it enables observers to map the wind currents, temperatures, and composition of the air around distant worlds, and the craft has been tasked with a heavy workload, studying the temperatures and size of Hot Jupiters, Super Earths, and smaller terrans. Spitzer may have found one of the smallest terrestrial planets, a rocky world two-thirds the size of Earth called UCF-1.01.

Spitzer was able to see exoplanets in the infrared part of the spectrum by chilling its sensors to very low temperatures using liquid helium. In 2009, its coolant ran out, but engineers reprogrammed the observatory to operate at higher temperatures. It is now embarking on its "Warm Mission," continuing its studies of the cosmos.

Hubble and Spitzer can analyze atmospheres around different worlds, not

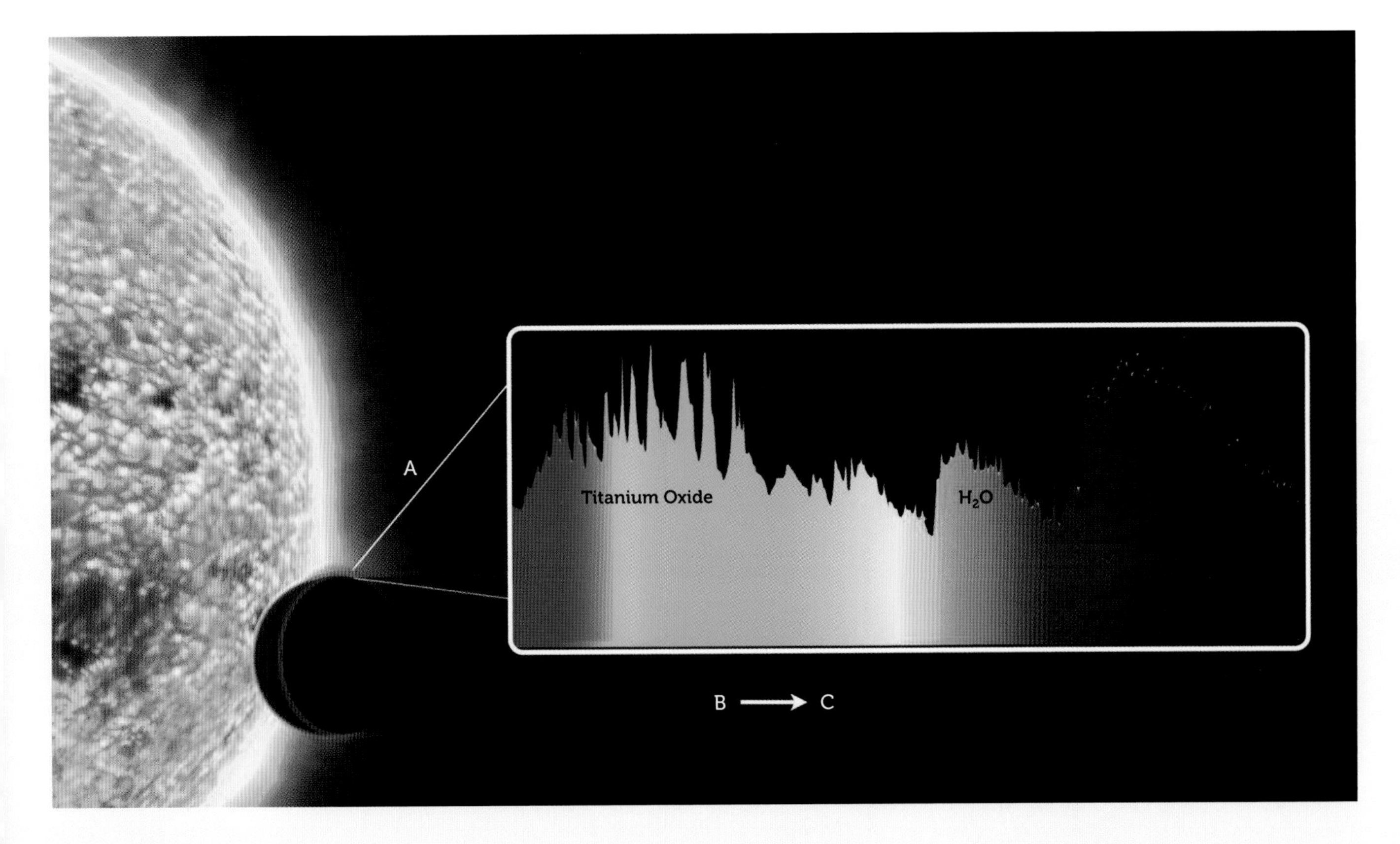

Above: *During a transit, starlight is partially blocked by a planet's atmosphere in a twilight ring around the planet, the terminator. As the starlight filters through the planet's atmosphere (A), its light provides clues to the makeup of the gases present (B–C).*

an easy task. When a planet is in transit in front of its star, the region between day and night, called the terminator, is the part of the planet that emits the spectrum of light from its atmosphere. NASA's Avi Mandell explains that, "When the planet goes in front of the star, you get absorption at the terminator and you see an increase in absorption at all wavelengths [colors] due to the atmosphere," but more of the light is absorbed by specific molecules. These shadowed bands in the starlight provide a fingerprint of the air surrounding the planet, but they are too dim to see through Earth's atmosphere. Our orbiting observatories have been able to tease out the gases in some of the exoplanets and astronomers continue to hone methods of sensing these critical clues to where life may be found.

The Kepler Observatory relied on a steady gaze for long periods to chart the dip in starlight caused by transits. Its impressive stability came from four pieces of equipment called reaction wheels—heavy disks that continually spin. When three or more are operating, their momentum keeps the craft in place and prevents it from drifting off target. But in the summer of 2012, one of Kepler's four reaction wheels froze up, and a year later a second one failed. This crippled the spacecraft, leaving it with no way to aim precisely, but clever flight engineers at Ball Aerospace figured out a way to use the craft's solar panels, orienting them in such a way that the pressure of sunlight stabilized the craft. In this way, the solar panels took the place of one of the failed wheels. Combined with the two remaining wheels, the solar panels assured that the mission could continue. Kepler's new lease on life was called K2, and it added dozens of planets to the Kepler Mission's discoveries.

SAILING THE COSMIC SEAS

A new armada of spacecraft—ambassadors from many spacefaring nations—is heading out into interplanetary space. Bristling with the latest technology, the flotilla is embarking on journeys designed to reveal more of the galaxy's exoplanets. In doing so, they will unveil more places where life may thrive in the cosmos.

TESS is an important and capable follow-on to Kepler. While Kepler was tasked with staring at one section of the star field, NASA's TESS is equipped to survey the entire sky. A Falcon 9 booster launched TESS in April 2018; during its first year of operation it studied the southern skies, gradually shifting to studies of the northern hemisphere in its second year.

TESS circles the Earth in an elliptical 13.7-day circuit. During each pass, the craft adds more and more "tiles" of the sky, assembling vertical strips that eventually will cover almost all of the galaxy. Where the segments overlap at the north and south poles, the craft will be able to observe for much longer periods of time. In these regions, planets following Earthlike orbits will be found, as these worlds cross their suns only once each year or so.

In 2021, the much-anticipated James Webb Space Telescope (JWST) will ascend to orbit. With a 6.5m (21ft) primary

JWST IS EXPECTED TO SEE PLANETS THAT ARE 10,000 TIMES FAINTER THAN THOSE SEEN BY KEPLER OR TESS.

Left: *NASA's Transiting Exoplanet Survey Satellite (TESS) will search for exoplanets throughout the entire sky.*

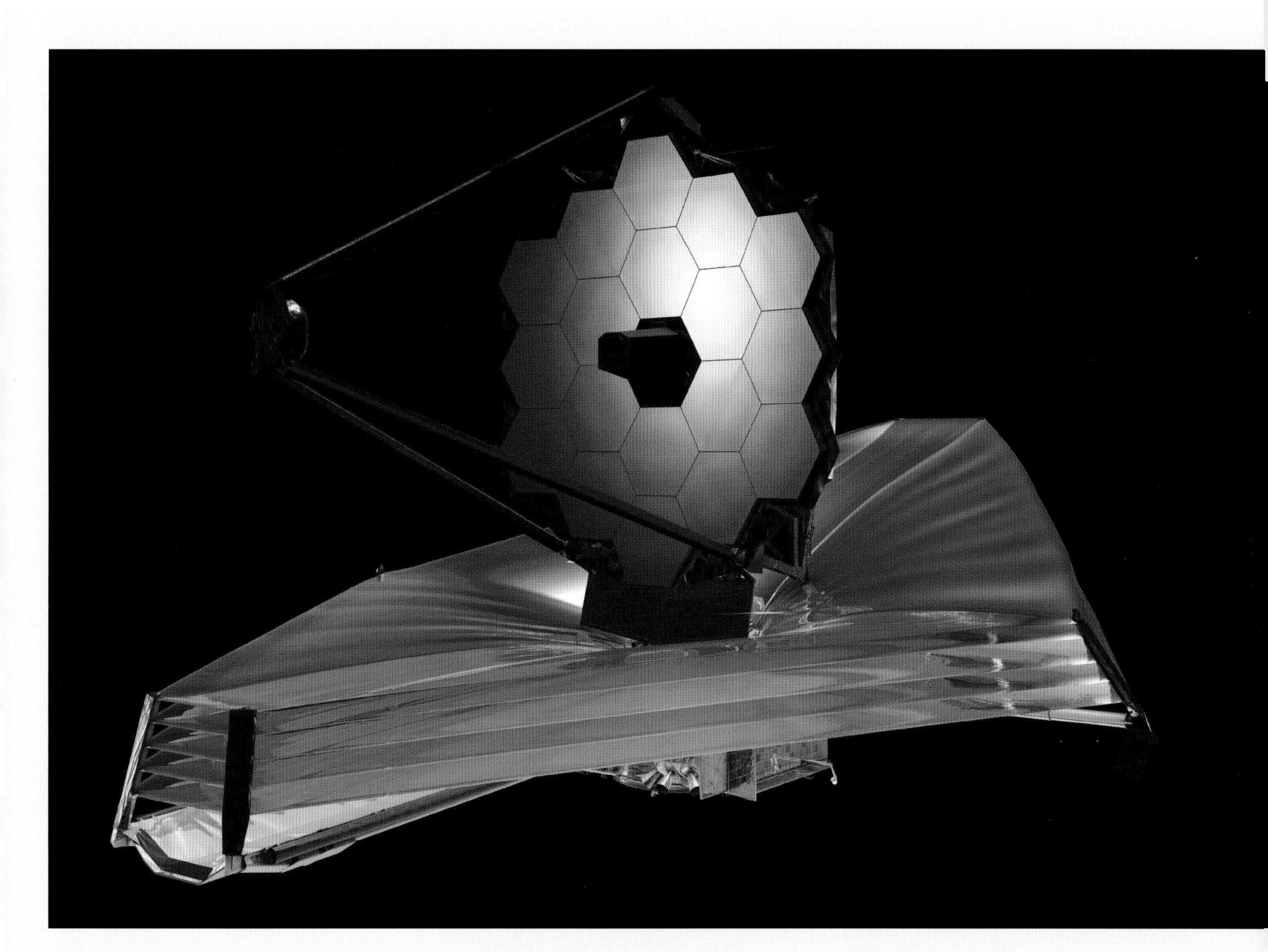

Above: *NASA's James Webb Space Telescope (JWST) ushers in the next generation of exoplanet research. Its 6.5m (21ft) primary mirror is made up of 18 hexagonal segments that will unfold to provide the best views yet of distant worlds.*

mirror, the craft is far larger than any space telescope yet launched, achieving 100 times the power of Hubble. It is designed to be the follow-on of Hubble (not specifically Kepler), as its mission will cover a vast array of topics, including the early stages of the big bang, exoplanet systems, and our own solar system's evolution. The craft has four separate imaging systems and a sun shield the size of a tennis court. JWST carries a set of coronagraphs, devices that block out sunlight so that the space very near them can be imaged. Using this light-blocker, JWST is expected to see planets that are 10,000 times fainter than those seen by Kepler or TESS, greatly expanding our list of known exoplanets.

The ESA's Gaia mission is in the process of measuring a billion stars over the course of a five-year period. The observatory uses astrometry to map five different factors, which include the position of the star at a specific reference time, the star's proper motion against the sky, and the parallax (how far the star appears to shift over one year due to the Earth's motion in its orbit from one

CHEOPS IS HELPING TO DETERMINE . . . WHETHER A PLANET IS GASEOUS LIKE NEPTUNE OR ROCKY LIKE THE TERRESTRIALS.

side of the Sun to the opposite side). If planets are present, the star's motion will be perturbed, so its motion against background stars will not match its predicted path.

While only a handful of exoplanets have been discovered using techniques that include astrometry from ground-based stations, several space telescopes have been able to use it to refine orbits of planetary systems discovered earlier with radial velocity. These include the ESA's Hipparcos and NASA/ESA's Hubble Space Telescope, as well as ground-based observatories like ESA's HARPS in Chile, the Hungarian Automated Telescope Networks (HATNet and HATSouth), and the Wide Angle Search for Planets (WASP), an international all-sky survey.

An exciting European space mission, headed by the ESA and the Swiss Space Office, has a different strategy. Rather than searching for new exoplanets, CHEOPS (CHaracterising ExOPlanets Satellite) will study known exoplanets to measure their diameters precisely. In this way, CHEOPS is helping to determine the accurate masses of planets that fall into the Super Earth and Neptune class. As astronomers lock down the masses of these worlds, they can better estimate their density and approximate composition. The measurements will help establish whether a planet is gaseous like Neptune or rocky, like the terrestrials. CHEOPS launched in December 2019.

Other upcoming proposed or planned missions in the search for exoplanets

Left: *The ESA-led CHEOPS mission analyzes the masses of exoplanets with the best accuracy in history.*

Above: *Upcoming exoplanet mission concepts include (from left to right): ARIEL, HabEx, and LUVOIR. With new technologies will come new discoveries.*

include the ESA's ARIEL (Atmospheric Remote-sensing Infrared Exoplanet Large-survey), which will study the atmosphere of a thousand exoplanets. NASA plans to follow up JWST with its Wide Field Infrared Survey Telescope (WFIRST), designed to determine how common solar systems like our own are and what factors regulate the habitability of planets. WFIRST makes use of gravitational microlensing. In proposal stage is LUVOIR, the Large UV/Optical/IR Surveyor, which—in addition to studying a wide array of cosmology—would examine many exoplanet types, from Hot Jupiters to small terrans, including those that may be inhabited even today. LUVOIR is one of four space telescopes vying for selection as NASA's next major spaceborne observatory.

The HabEx (Habitable Exoplanet Observatory) space telescope would search for and image Earth-size habitable exoplanets in the habitable zones of their stars by using a free-flying star shade to block out the glare from the stars. Another, the Lynx X-ray Observatory, would detect and study extremely faint exoplanets, many of them Earth-class. The Origins Space Telescope would focus on the formation of other suns across the Milky Way.

According to a NASA release of the missions under consideration: "We'll have all-new textbooks." But will it be enough to find life in the galaxy?

WHAT ARE WE FINDING?

AN OVERVIEW OF NEW WORLDS

Armed with Kepler's powerful transit observations, and with advances in other planet-hunting techniques using other observatories, astronomers have been able to pin down a multitude of planets orbiting distant suns, and they've been determining something about their nature as well. It hasn't been easy, says NASA Goddard Space Flight Center's Avi Mandell. "When you detect a planet with the transit method, then you only get the radius. You get nothing about the internal composition or density. You have a size where the planet blocks out the starlight. But it's even unclear whether you're seeing the solid surface of the planet or some thick atmosphere extending around it. So you often have no idea except for a circular disk that's blocking out the light." But once researchers estimate the area of the star covered by the planet, they begin to add other parts of the puzzle, and one of the most important parts is mass. Using techniques like radial velocity astronomers can get at least some general constraints on the planet's mass. But Mandell warns that, "You have to know a lot about your star because it's really about a relation to the star's mass. If you know the star's mass, then you get a good mass constraint on the planet."

The process is far more difficult with stars that vary in luminosity, as we have seen with M dwarf flare stars. Other stars behave badly for transit techniques, too, among them the red giants, huge stars that vary in brightness at irregular intervals.

The random brightening and dimming of light are not the only problems facing planet hunters. Another complicating factor is the distance of a planet to its star. Planets in tight orbits around small M dwarfs may cross the face of their sun again and again in a matter of weeks or even days, making them easy to spot in the ebb and flow of starlight. But planets orbiting at a distance comparable to that of Earth from the Sun will not repeat their crossing for a year, making them far more difficult to track.

Despite the challenges, Kepler netted 3,924 exoplanets within its first few years (see "The numbers game" feature on pages 88–89). But what of Earthlike worlds? Some 156 of Kepler's initial 4,000 appear to be rocky and close to the Earth in size. Some lie in habitable zones, while others smolder close to their suns or freeze in distant, sterile orbits. And while these are the targets of our search for Earth 2.0, Kepler unveiled a host of surprising exoplanet discoveries, their natures as astounding as their numbers. And the architectures of their solar systems were a shock, quite different from the arrangement of our own planetary system.

One of the more intriguing revelations is the planet WASP-121 b. WASP-121 itself is a star quite similar to our own Sun. It has a comparable mass and diameter, though it is slightly hotter and brighter. Circling the star in a tight orbit lasting just 1.27 days is WASP-121 b, a Hot Jupiter. The planet is so close to its star that its

156 OF KEPLER'S INITIAL 4,000 APPEAR TO BE ROCKY AND CLOSE TO THE EARTH IN SIZE. SOME LIE IN HABITABLE ZONES . . .

Right: *Atmosphere streams from a Hot Jupiter, seen from an airless moon. Solar wind from the planet's nearby star strips away the gases. Eventually, only a rocky core will remain.*

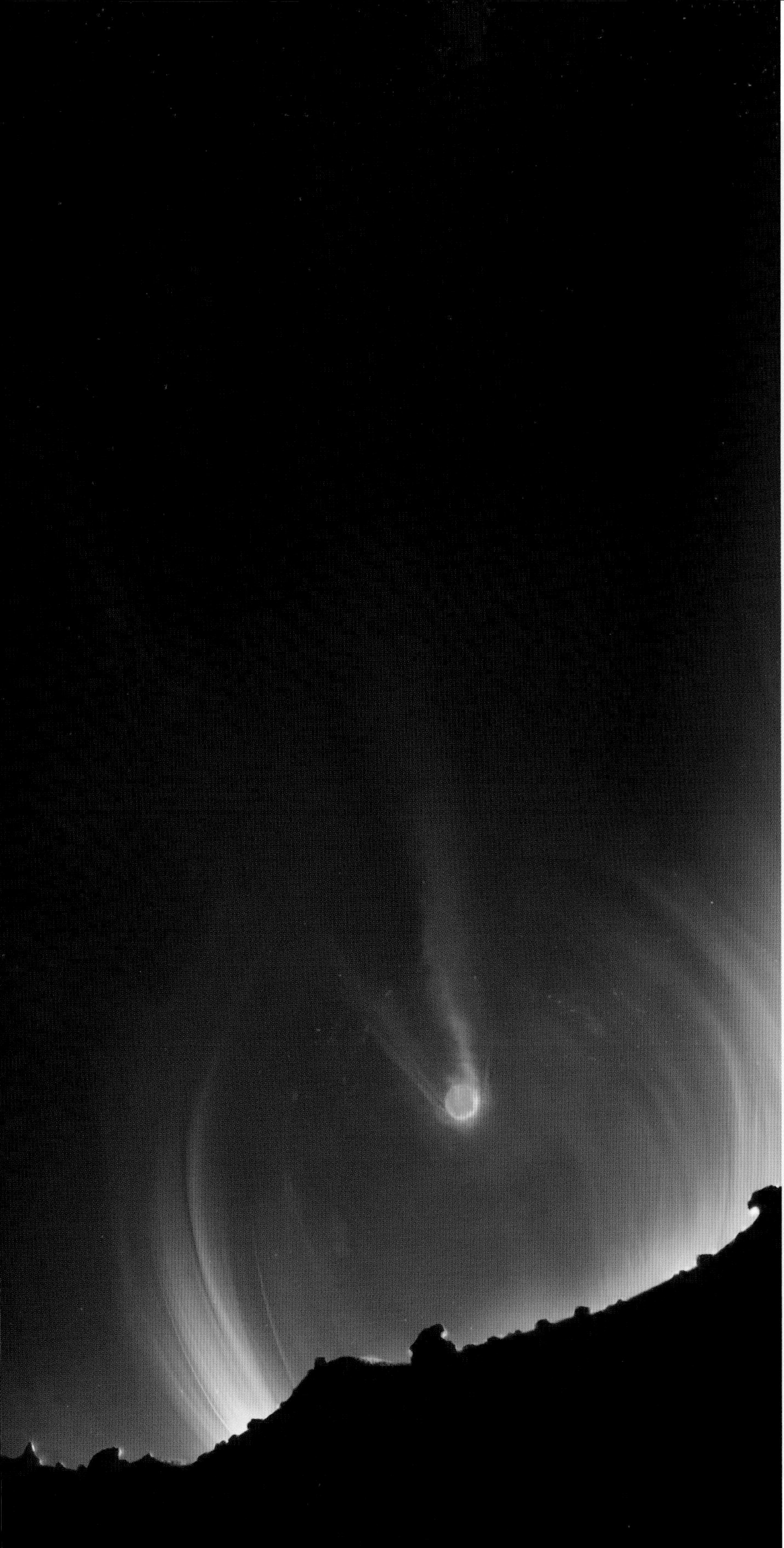

atmosphere is streaming away. Within that atmosphere, Hubble has detected iron and magnesium. These gases have been found in other Hot Jupiters, but those planets have been cool enough that the magnesium and iron condense into clouds, while WASP-121 b is so hot that the metals are being stripped away in the furious solar wind of nearby WASP-121. Its upper atmosphere blazes at 2,538°C (4,600°F), 10 times hotter than any other planet yet found, and it orbits so close to its sun that the planet likely has an elongated shape, much like a football. WASP-121 b is also the first exoplanet where water has been confirmed in its atmosphere.

Another Hot Jupiter lies just 5,745,000km (3,569,780 miles) from its star, 25 times closer than the Earth to the Sun. HAT-P-12 b orbits the star HAT-P-12, a K-type star (more orange and cooler than our Sun). HAT-P-12 lies some 468 light-years from our own solar system and circles its star in a tight orbit, racing along a complete circuit over the course of just 3.2 days. Although gaseous, its bands are probably not parallel like those of Jupiter. Instead clouds may stream radially from the point on the planet facing the star directly, as HAT-P-12 b is tidally locked, keeping the same face toward the star. Heat will move from the central dayside away toward the night, setting up fierce winds that radiate like spokes on a wheel. While the planet is about the same size as Jupiter, its mass is only 11 times that of Earth, or less than

While Kepler has found or studied many bizarre worlds, among the most exotic is Kepler-2 b, also known as HAT-P-7b. The planet was one of the first Hot Jupiters found. Shortly after its discovery by the HATNet project, Kepler ascertained that the giant planet is larger and denser than Jupiter. With an orbit just 5.5 million km (3.4 million miles) from its star, the planet's cloud tops roast at 2,450°C (4,440°F). Computer models suggest that ruby and sapphire dust condense from the clouds on the nightside, creating a truly alien weather system.

Above: *Twilight on a Hot Jupiter: HAT-P-7b also has a nightside. Incandescent clouds glow with the heat carried by winds from the dayside. Ruby and sapphire vapors shimmer in the sky. The furious dance of aurorae appears as curtains that mark radiation fields raining down from the star on the other side of the planet.*

Left: *Circling a Sun-like star 700 light-years from Earth, the hot gas giant TrES-2b may be the darkest planet in existence. Seen from a nearby moon, the planet would practically vanish against the starry sky.*

Above: *Kepler-10 b, one of the first rocky exoplanets found, smolders in the heat of its nearby star. The planet circles its Sun-size star at just one-twentieth the distance of Mercury to the Sun.*

Another strange revelation for astronomers is the Jupiter-size world TrES-2b. Though not discovered by Kepler, the observatory pinned down the planet's mass. Kepler was also able to determine just how black TrES-2b is. The gas giant has been called "the darkest world in the galaxy." It is less reflective than coal dust, and even darker than Kepler-2 b. The reason is unknown, but may be due to a lack of bright clouds seen so often on the gas giants of our solar system, or light-absorbing vapors like titanium, sodium, or potassium. The planet orbits just over 5 million km (3 million miles) from its Sun-like star.

A third Kepler target is Kepler-10 b, a lava world. Half again as large as the Earth, this grilled globe orbits so close to its star that surface temperatures hover at 1,560°C (2,840°F). It is a rocky, molten world and probably has retained enough internal heat to suffer widespread volcanism today.

Above: *In the early days of planet-hunting, astronomers assumed that planets could not form within systems of multiple stars. Kepler and other observatories have revealed that this is not the case. Here, artist and astronomer Dirk Terrell envisions the view from a moon of the gas giant in the four-star system of Kepler 64.*

Left: *Sunsets in a multiple star system would be spectacular affairs indeed. Often, different types of stars are found together. In this imaginary view, a G-type, Sunlike star at left compliments the cool light from a white dwarf at right.*

a quarter that of Jupiter, which makes HAT-P-12 b the least dense gas giant found so far. The planet was discovered by the Harvard/Smithsonian HATNet project using the transit technique.

Like most planets that orbit close to their stars, the gas giant is tidally locked, always pointing the same face toward its sun. This spin, slower than that of Jupiter or Saturn, likely results in belts that are far more subtle. Jupiter's well-defined belts partially result from its short day: the huge planet spins around in just under 10 hours. Gold-banded Saturn—nearly as large—turns once each 10 hours, 42 minutes. But since HAT-P-12 b is tidally locked, its daily spin matches its circuit around the star, making a turn once every 3.2 days. This comparatively leisurely rotation will not split the atmosphere into the kind of tight bands seen on the gas giants of our system. In fact, says NASA's Avi Mandell, the circulation of the huge planet's clouds is much more intriguing. "You have these equatorial jets. You get these amazing chevron-shaped forms like a large arrow pointing in the way of the rotation, with lobes coming out from the jets. So the planet is heated at the equator and then these cells circulate material poleward." The Hot Jupiters of the cosmos may not look much like our own Jupiter at all.

STAR WARS SUNSETS

Despite the elegant double sunset on *Star Wars* planet Tatooine, many astronomers calculated that planetary systems would be unstable if they orbited in a multiple-star system. The majority of star systems in the galaxy have multiple stars; our Sun is an exception, not the rule. In fact, the nearest star system to us, Alpha Centauri, is a three-star system consisting of a Sun-like star, a cooler orange K star, and a tiny red dwarf. The red dwarf, Proxima Centauri, is the closest star to our solar system. If multiple-star systems could not sustain

THE NUMBERS GAME

Astronomers have now located nearly 5,000 confirmed or candidate exoplanets. Of those, the largest Jupiter-like globes sign in at 1,213. Some 1,665 are smaller gaseous worlds roughly the size of Neptune. A total of 878 could be considered Super Earths, their sizes falling between those of Neptune and Earth. Their forms probably vary from large balls of rock to worlds of water vapor and superdense gases. One hundred fifty-six are seen as resembling Earth in size and mass (but perhaps not in nature), while a dozen or so of the total are classified as "unknown," their true character not yet understood.

Researchers estimate that there are hundreds of billions of planets in our Milky Way galaxy. It's a big number to throw around—how do they know? The number of stars in the Milky Way is surprisingly difficult to estimate. Our galaxy is 100,000 light-years across, with a central bulge and four spiral arms. The two largest arms of stars are called Perseus and Sagittarius. There are two smaller spurs, one of which is called the Orion Arm. It is here, in a mundane backwater of the galaxy, that the Sun and her planets reside. Based on star types, their average masses, and estimates of the Milky Way's mass, researchers estimate that our galaxy contains approximately 200 billion stars. The Kepler spacecraft has observed about 145,000 stars. If this assortment of stars is representative of the entire population of suns within the galaxy, and most of those stars have multiple planetary systems, simple math tells us that there should be hundreds of billions of planets floating around the Milky Way neighborhood. And instead of all looking like the ones we're familiar with in our solar system, they seem to come in all shapes and sizes. The exoplanetary zoo is expanding and changing, and with it, so is the search for life within that menagerie.

THE EXOPLANETARY ZOO IS CHANGING, AND WITH IT, SO IS THE SEARCH FOR LIFE WITHIN THAT MENAGERIE.

EXOPLANETS BY THE NUMBERS

Confirmed exoplanets revealed by Kepler	2,345
Kepler candidates awaiting confirmation	2,420
K2 confirmed planets	389
K2 candidates awaiting confirmation	892

Right: *This Hubble image of the galaxy NGC 6744 gives us a sense of the myriad stars in our own galaxy. Of the approximately 200 billion suns in the Milky Way, how many have life-sustaining planets?*

NOT ALL EXOPLANETS ARE DISCOVERED BY PROFESSIONAL SCIENTISTS. HIGH SCHOOL INTERN WOLF CUKIER . . . DISCOVERED AN EXOPLANET IN ORBIT AROUND TWO STARS.

Left: *Kepler-35 b orbits two stars. The double suns are similar to our Sun, but slightly smaller and cooler. Kepler-35 b is the size of Saturn, but undoubtedly of a very different nature as it resides in a hot region around its stars.*

THE TWO SUN-LIKE G STARS ORBIT EACH OTHER AT A DISTANCE OF ONLY 30 MILLION KM . . . THE PLANET KEPLER-35 B KEEPS ITS DISTANCE . . .

stable planetary systems, that would leave out a great number of star systems where we might find planets, let alone life—but eventually, planets were found orbiting such systems. One such star grouping is known as Gliese 667, a triple-star system. One member, Gliese 667C, has at least six planets, three of which orbit within its habitable zone. Gliese 667C c is a Super Earth 1.8 times as far across as the Earth, circling its star every 28 Earth days. Gliese 667C e and f both measure 1.5 times Earth's diameter. Gliese 667C e makes its circuit every 39 days, while f's year comes in at 62 days.

Another planet loiters in the multiple-star system of Kepler-35. Sibling suns of the gas giant Kepler-35 b circle around each other every 21 days. The two Sun-like G stars orbit each other at a distance of only 30 million km (19 million miles), half the distance from Mercury to the Sun. The planet Kepler-35 b keeps its distance from this stellar dance, orbiting the two stars every 131 days from a distance of about 0.6 AU. The planet's orbit is not circular because it circles two stars instead of one; therefore it travels in a somewhat chaotic oval track. Kepler-35 b is proof that multiple-star systems can sustain planetary families—but how habitable those systems are is still up for debate. For its part, Kepler-35 b orbits well inside the habitable zone of its double suns.

Note that not all exoplanets are discovered by professional scientists. High school student Wolf Cukier did an internship at NASA Goddard. Tasked with sifting through transit readings from the TESS spacecraft, he discovered an exoplanet in orbit around two stars. While other planets have been found circling double stars in the past, this was the first discovery using data from the TESS spacecraft. TOI 1338 is a two-star system with a Sun-like star orbited by a red dwarf. TOI 1338 b crosses the face of both stars, but its transits across the fainter red dwarf are too subtle to detect. Still, light from the double-star system drops in the familiar pattern of an orbiting exoplanet. The planet itself is large, between the size of Neptune and Saturn. Its roughly 94-day orbit puts the planet closer to its double suns than the habitable zone.

KEPLER FINDS SOME STONY WORLDS

The star cataloged as HR8832 (also designated HD 219134 or Gliese 892) is an orange in-between sun, a K-type star that weighs in somewhere between a red dwarf and a G-type star like our own Sun. With a mass eight-tenths of the Sun's, the star is cooler, simmering at 4,425°C (8,000°F) (compared to the Sun's toasty 5,505°C / 9,941°F). HR8832 has a radius 0.78 that of the Sun. (One of the closest similar K-type stars to Earth is Alpha Centauri B.)

HR8832 is not alone, as the star has at least five exoplanet companions with a range of sizes, from a Saturn-size world to the innermost rocky world, HR8832 b. HR8832 b's size falls between that of the

Left and right: *The skies of a planet orbiting a double star might be quite different from our Earth's firmament, with its ordered sunrises and sunsets. Here, a hypothetical planet in an elliptical orbit around a double-star system endures extreme temperature swings as it makes its way around its orbit outside of two suns.*

Earth and the ice giants, qualifying it as a Super Earth. It is 1.6 times as far across as our world, but it's a heavyweight, weighing in with a mass equal to 4.75 Earths.

HR8832 b's orbit lies painfully close to its star; the planet circles at a scant 5,837,000km (3,627,000 miles), making its year a short 3.1 Earth days. Discovered by the radial velocity technique, observers have since been able to determine that the planet has an atmosphere, and that its surface fries at 742°C (1,368°F). Because of its nearness to the star, its surface cannot be covered by global ocean, but is probably rocky. The atmosphere may be thin and clear or may be a dense covering of opaque water vapor clouds. Its terrestrial nature and size make it likely that the planet is geologically active and may host global volcanism. Another rocky Super Earth orbits twice as far from the star. This supersize terran weighs in at 2.7 times the mass of Earth. Still farther out, a third Super Earth orbits in a 47-day circuit;

THE ARCHITECTURE OF THE HR8832 PLANETARY SYSTEM MIMICS THAT OF OUR OWN SOLAR SYSTEM, WITH ITS DENSE AND ROCKY INNER PLANETS . . .

this one is nine times the diameter of Earth. Two more Super Earths may orbit outside of these, yet to be confirmed.

Some 2.1 AU beyond lies a planet just smaller than Saturn. HR883 2h has a mass 108 times that of the Earth and is thought to be a gas giant. While the planet itself has no solid surface, atmospheric conditions on moons circling it would make it possible for liquid ammonia to serve a similar role as water does in Earth's environment, raining or snowing from the skies and pooling in ammonia lakes and seas. (Methane serves the same role on Saturn's moon Titan, where it condenses as clouds and rains into extensive lakes and seas.) The architecture of the HR8832 planetary system mimics that of our own solar system, with its dense and rocky inner planets and probably two gas giants in an outer realm (planets e and h). Confirmation of the existence of planets f and g requires further study.

©M. Carroll

Right: *Many of the planets discovered by Kepler and other observatories circle their stars in close orbits, like the rocky planet Corot-7 b. Likely a terrestrial planet, Corot-7 b circles its Sun-like star once every 20 hours. It was found by the French planet-hunting CoRoT mission.*

Left: *The gas giant HR8832 h looms in the sky over a large, atmosphere-enshrouded moon. Its orbit lies in a zone far enough from the star to host frozen and liquid ammonia on the surfaces of any nearby moons. This view of an ammonia "waterfall" envisions an atmosphere tinted green by the presence of methane.*

COOKING CELESTIALS

Many of the planets discovered by Kepler and other observatories circle their stars in close orbits. This is covenient, as it makes them easier to find, as they cross the faces of their suns often. One such world is Corot-7 b. It is half the size of Earth, but thought to be a terran planet rather than a gaseous one. The French CoRoT observatory—the first spacecraft earmarked specifically for exoplanet research—unveiled the planet after studies leading up to 2009. Corot-7 b races around its star once each 20 hours, 29 minutes. It is not the only planet circling its sun: another Super Earth, Corot-7 c, orbits farther out, taking just 3.7 days. Neither are promising targets in our search for life as temperatures on both would vaporize any biological material, or even some metals!

Corot-7 b seems to be a hellish place, but it has nothing on a molten planet orbiting the star 55 Cancri A, a G-type star similar to the Sun. With a mass nearly nine times that of Earth, this Super Earth is nearly twice as far across as our own world. Its orbit is so close to its star that its sun-facing hemisphere cooks at 1,700°C (3,100°F). It is so hot, in fact, that the surface is hot enough to vaporize rock. Hubble was able to detect hydrogen, helium, and hydrogen cyanide, but no water. Across its skies drift clouds of sapphires, says astrophysicist Caroline Dorn of the University of Zurich. "I imagine 55 Cancri e to be a magma ocean world. In other words, rocks on the surface are so hot that they are molten and they partially evaporate, since temperatures on the dayside of this planet are around 2,700 K . . . these are temperatures at which rock simply evaporates. Therefore, there is likely a tiny atmosphere (at least 10 times less than for Earth) made from mineral-rich gases." Some modelers assert that both sapphire and ruby clouds ride the furious winds of the planet. The clouds may well be made of corundum, which forms these semiprecious gems. "It's a crazy hot world," Dorn says, "where we can imagine that no water and no Neptune-like atmosphere can exist."

Above: *The molten rock surface of 55 Cancri e evaporates into a thin atmosphere containing rubies and sapphires. The planet's blistering surface maintains a temperature at the vaporizing point of basalt.*

WASP-49 b is a broiling cousin to Saturn. It is nearly the same mass, but the Hot Jupiter features something not found elsewhere. The giant planet appears to be embedded in a vast cloud of sodium. We do know of another planet surrounded by sodium: Jupiter. The cause of Jupiter's gaseous cocoon is its famous volcanic moon, Io, a tiny globe about the size of Earth's Moon. While the Earth's Moon is geologically quiet and battered with craters, Io's face has been resurfaced by hundreds of active volcanoes. Some resemble the basaltic volcanoes of Earth, mountains that erupt molten rock, but others are gaseous plumes of sulfur dioxide and other gases shooting hundreds of kilometers into the sky. These plumes pump the entire Jupiter system full of sulfur. Io's activity is due to a powerful force called tidal heating: the gravity of Jupiter and its other large moons plays a cosmic tug of war with Io, pushing and pulling it to the point that its surface rises and falls some 50m (160ft) each day. This movement within the moon causes its interior to heat up, and that heat comes out as some of the most spectacular volcanoes in our solar system. Some observers think it likely that WASP-49 b has a similar moon spewing out sulfurous volcanic plumes.

"You could imagine these massive plumes blowing not into an atmosphere, but into a vacuum and then swept away by the solar wind," says NASA's Avi Mandell. "You might have volcanoes on the nightside because of tidal heating. You get what's called the heat-pipe structure, which is very similar to Io's. Conceivably, you could have hot spots all over the planet."

Worlds like 55 Cancri e and WASP-49 b open a window into a solar system's

Above: *Jupiter's moon Io (left) bathes the gas giant and its moons in a cloud of sulfur. A similar situation, seen at right, may be occurring in the WASP-49 b system.*

past. Stars like our own Sun are initially cocooned in disks of dust and gas, with volatiles—like water—freezing into ice in the outer parts of the cloud. Rocky, terrestrial planets condense from pebbles, solid boulders, and asteroids left behind as the solar wind disperses the lighter material in the disk. In solar systems like ours, these planetary building blocks form in the inner part of the cloud, and include iron and magnesium. The process leads to terrestrial planets with iron cores. Many of the Super Earths found so far formed under such conditions, but some planets condense closer in, where temperatures are far hotter. In the innermost stellar neighborhood, some elements are still gases, not able to cool into solid material for planet building. Their compositions are very different, ruled by calcium and aluminum along with silicon and magnesium. These high-temperature condensates lead to planets of very different makeups, worlds whose atmospheres, crusts, and cores add up to new classes of Super Earths. Since their cores lack iron, they can have no magnetic field (which is generated by currents in molten metals surrounding the solid core), and as with 55 Cancri e, their atmospheres will be quite alien and unfamiliar.

3

CIRCLING A RED DWARF

Kepler and other observatories have given us a taste of worlds out there. But to truly understand the possibilities of finding life among the exoplanets, we must dig in further. We begin with planets orbiting the most common stars in the universe, the red dwarfs—those older, colder stars, smaller than our Sun, which have a nasty tendency to jettison deadly flares (hence also called "flare stars"). In the search for life in the galaxy, scientists estimate 20–30 percent of all exoplanets are Earthlike in size and temperature. That's tens of billions in our Milky Way galaxy alone. Researchers at Puerto Rico's Arecibo Observatory continually update their list of worlds in the Habitable Exoplanets Catalog. As of September 2019, that list is up to 52, from a total of over 4,000 confirmed planets: 33 are "superterran," falling between Earth and Neptune in size (Super Earths); 18 are roughly Earth-size; and another one is "subterran," measuring a similar diameter to Mars (see the "The 'A-list' of exoplanets" feature on pages 110–111). Though not Earth twins, these red dwarf worlds could be (distant) cousins. Might they host life?

Left: *If Proxima b has a mass several times that of Earth, it will qualify as a mysterious Super Earth, a mix of terrestrial planet and gaseous world. Here, we see Proxima b as a Super Earth viewed from a nearby large moon, itself a candidate for life. Large-mass planets like this version of Proxima b are considered less likely than smaller ones.*

LIFE NEAR A RED DWARF STAR

The sun never sets on Teegarden b—that is, if you're on the sunlit side. Like so many of the planets discovered in orbit around red dwarf suns, Teegarden b must be tidally locked, with one hemisphere always facing its parent star. The reason has to do with gravity: Teegarden b is so close to its sun that the gravity of the star and planet have locked each other in a grip where the planet turns exactly once for each revolution around its star. The planet's somewhat elegant-sounding appellation comes from the name of its sun, Teegarden, an M-type red dwarf star about 176,700km (110,000 miles) across, just one-third larger than Jupiter. The star was named after its discoverer, astrophysicist Bonnard Teegarden, who led a team that studied data from telescopes of the Near-Earth Asteroid Tracking program. Although designed to verify asteroids that cross the Earth's orbit, their search also captured the light from background stars, including Teegarden. The dim star's existence was buried within the data for several years before the team took a look. Teegarden is the 24th closest star to our Sun.

Teegarden's star counts itself in the majority of suns in the Milky Way. Nine out of 10 stars in the night sky are invisible from your backyard (unless you have a very powerful telescope), and the red dwarfs—or M stars—akin to Teegarden burn coolly, dimly, and slowly. Some are among the most ancient objects in the Milky Way, and many are as small as Jupiter.

Red dwarf expert Allison Youngblood at the Laboratory for Atmospheric and Space Physics (University of Colorado Boulder) suggests that the M stars offer unique possibilities. "Why red dwarfs? Because it's easy to find planets around them, and they're abundant and nearby. They make up most of the stars. If we have a shot at finding an Earthlike exoplanet, it's probably going to be around an M-type star. Also, the planets are closer. The habitable zone is much closer to the star, which makes the planet easier to detect; it makes their signals larger." The quality of the starlight is significant in the search for life-friendly planets, Youngblood says. "Most of the radiation is in infrared [heat]. It makes things like habitability interesting because you're so close in to the star."

As the most common stars in the galaxy, red dwarfs host a wide variety of planets, from tidally locked "eyeball" planets to ice and gas giants, including such exotic globes as Hot Jupiters and Super Earth water worlds. Because planetary systems circling red dwarfs are closely packed around their small stars, the skies of these worlds are

Right: *The rocky terran Gliese 677C c orbits a red dwarf star, part of a triple-star system. The exoplanet was found by the telltale shift in starlight that indicated a planet pushing and pulling on the star Gliese 667C. Recent calculations show the planet to be in its star's habitable zone. The stars Gliese 667A and B glow to the right.*

NINE OUT OF TEN STARS IN THE NIGHT SKY ARE INVISIBLE FROM YOUR BACK YARD . . .

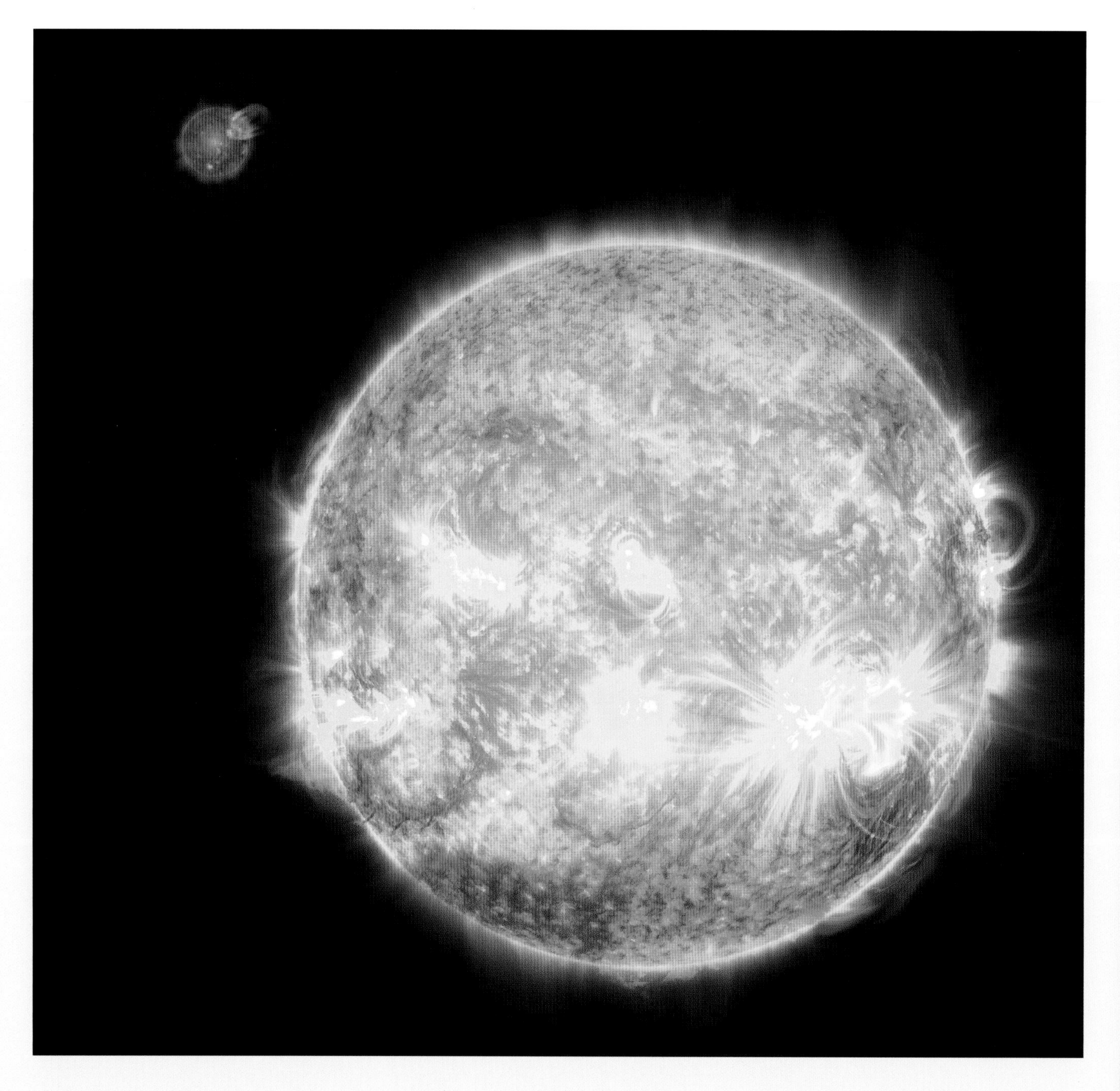

often bejeweled with nearby planets, companions periodically passing by at close range. Some of these encounters would yield spectacular views, with the passing planet many times the visual size of a full Moon in Earth's sky. But the close proximity to the star itself comes at a price: such an orbit locks one face of a planet toward the star, placing the nightside in eternal darkness (see "Eyeballs: Bizarre Planets with 20/20 Vision" on page 112).

The planets of Teegarden are a good case study of worlds in a red dwarf

Above: *The red dwarf Teegarden's star, at upper left, compared to the Sun. The M star is about one-tenth the size of the Sun, just a bit larger than Jupiter, but much more massive.*

Right: *Red dwarf stars habitually cast out powerful flares. These high-radiation fountains can be comparable to the most powerful stellar ejections from our Sun, but they occur far more frequently.*

Below: *So far, only one discovered exoplanet falls into the subterran category, a planet about the size of Mars. Even within the center of a habitable zone, such a small planet may not be able to hold on to a substantial atmosphere to support surface water. But oceans may well survive under the surface, and life might thrive there or in polar caps that retain liquid water on the margins.*

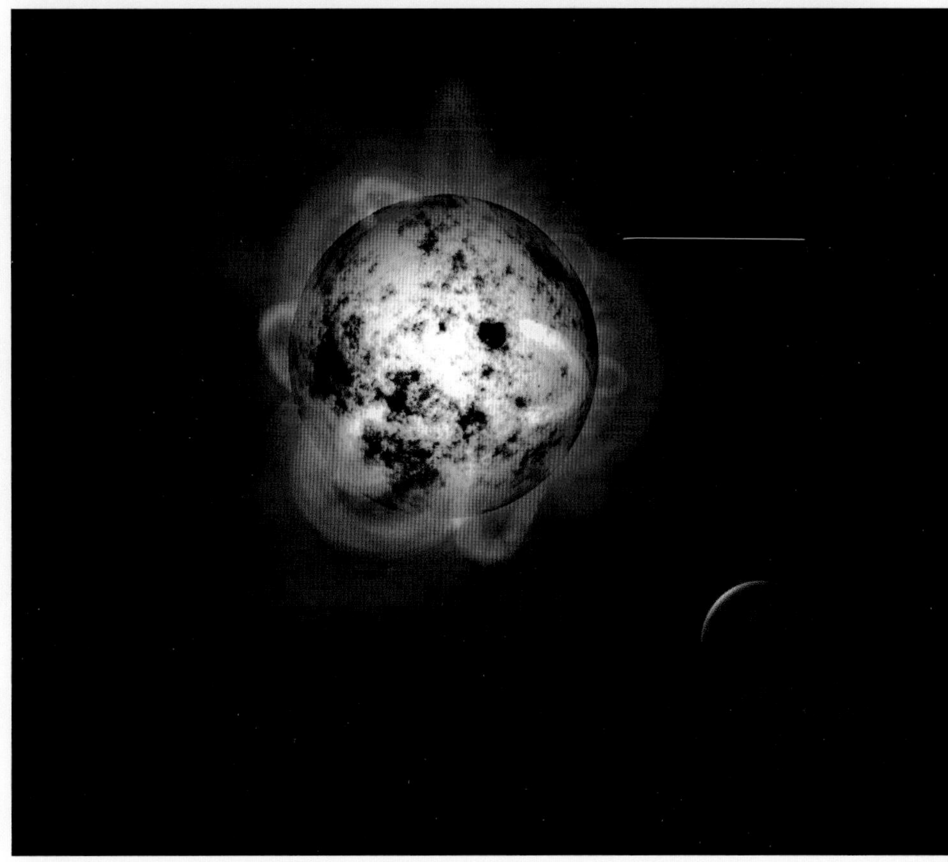

Right: *A red dwarf's strong magnetic fields send glowing arcs of material into near space. An exoplanet lurks nearby. Although a red dwarf's habitable zone lies within range of the star's dangerous radiation outbursts, a strong magnetosphere might shelter the surface, enabling life to take hold and flourish.*

system. In our search for life in the galaxy, candidates in a red dwarf system offer sound targets as many occupy the habitable zones of their cool stars. They must orbit their suns at short distances due to the coolness of their star. This makes their year—the length of time the planet completes a circle around its star—very short compared to our own home world. The Earth's year of 365 days is determined by its distance to the Sun, but our Sun is far hotter than the typical red dwarf, so its habitable zone is farther away (see "Living the good

Above: *Basking in the light of its cool sun, Teegarden b orbits within the habitable zone. If the planet were able to hold on to substantial atmosphere and water, it would be a prime candidate for a benign environment that might host life. Here, we see an active water cycle on the planet, its sky tinted red by its red dwarf sun.*

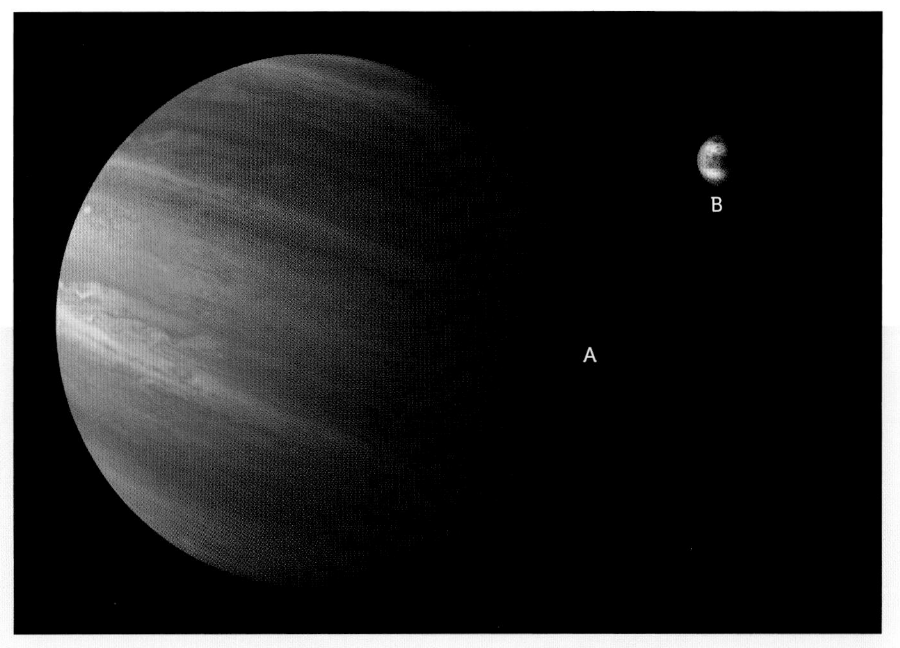

Above right: *Hot Jupiters appear to be common around red dwarf stars. Here, the rocky Super Earth Kepler-452 b (B) is compared to 51 Pegasi b (A), the first Hot Jupiter (and first confirmed exoplanet) ever found. The two exoplanets were discovered 20 years apart, in 2015 and 1995, respectively.*

life (in the Goldilocks zone)" feature on pages 24–25). Planets of cool M stars like Teegarden orbit so closely that their years are on the order of days or weeks rather than months or years.

Red dwarfs are strange stars. Many have diameters close to that of Jupiter. While the Earth's Sun burns at 5,500°C (9,930°F), Teegarden smolders at a temperature of 2,365°C (4,290°F). Like other stars of its kind, its low temperature makes the M star reddish, lending it the nickname red dwarf. And while it is dim in the visible spectrum, M-star planetary systems don't need as much energy output to heat the habitable region, says NASA's Ravi Kopparapu. "You don't need much stellar radiation from M stars, because they are brighter in the infrared part of the spectrum, and water is a good absorber." It doesn't take much heat to warm a planet around an M star, either. As a planet moves closer to the star, if that planet has water on it, the water slowly evaporates and goes into the atmosphere. But since water is a good infrared absorber and the star itself is a good infrared emitter, Kopparapu says, "it's a match made in heaven. The planet is ready to take the heat, and the star is ready to give it. The planet gets warmer and warmer, so you don't have to push the planet much closer to the star." In other words, the habitable zone of an M star can remain farther away from the star and its deadly flares, all because of the type of light given off by the red dwarf.

Teegarden has at least two planets, labeled b and c. The planets do not transit their star as seen from our standpoint, so could not be sensed by the Kepler Space Telescope. But because the star is so close to the Earth, just 12.6 light-years away, its motion could be studied. Teegarden's little solar system was revealed by the radial velocity technique. Shifts in Teegarden's starlight showed a wobble in the star, the telltale sign that planets were nearby, pulling on their parent star as they circled it. With over 200 observations combined, researchers were able to pin down a rough idea of the size, mass, and location of the planets.

Astronomers estimate that Teegarden b, the innermost planet, has a 60 percent chance of hosting an environment that

holds liquid water. Temperature estimates range from 0–50°C (32–122°F), but the planet's daytime temperatures may well hover around 28°C (82°F) if the atmosphere is similar in extent to our own. Teegarden b's ESI is the highest of any world yet found. The planet is only slightly heavier than the Earth, so is assumed to be essentially the same diameter.

Teegarden b's sibling, Teegarden c, is not such a temperate place. This second planet of the system is farther out and slightly more massive, with temperatures that probably never reach over −47°C (−52°F), similar to those on Mars. The planet's mass is at least equal to Earth's and it is likely a terran in nature, although it may be covered by frozen global oceans if its size is substantially larger.

Teegarden b checks in at an impressive 0.93 on the ESI, the highest yet found. It is a strange, orange-tinted world with a year lasting only five Earth days. The swollen sun—spanning more than 10 times as far across as our Sun appears in the skies of Earth-bound observers—continually shines like a sullen ember, occasionally lighting up the sky with exploding flares.

The more distant Teegarden c has a 0.69 ESI, with only a 3 percent chance of having a warm surface temperature. This is partly due to its likely thin atmosphere.

In fact, a lack of atmosphere may be a problem for any life potential among the planets of Teegarden, as well as among all terrestrial planets near red dwarf stars. NASA's Avi Mandell warns that the smaller, rocky planets in a red dwarf's habitable zone may end up with little or no atmosphere at all because of their proximity to the star. "The problem with rocky planets is that your atmosphere is often going to be driven off. There are a

MANY ROCKY, EARTH-SIZED WORLDS ORBIT WITHIN THEIR STAR'S HABITABLE ZONE, BUT THEIR SUNS POUR OUT SEARING FLARES AND DEADLY X-RAYS.

Left: *A shallow sea lingers beneath an icy crust on Teegarden c. The planet lies at the outer edge of the habitable zone. Teegarden's feeble light tints the snow in purples and reds.*

Below: *An alternative view of Teegarden c shows the planet within the habitable zone, but nearly devoid of atmosphere. Early on, its red dwarf sun may have stripped away most of its air. Any volatiles would remain as scatterings of ice, sheltered in the shadows. If the estimates of some researchers are correct, this desolation may be the fate of most planets orbiting M stars. In the background, Teegarden b floats in the glare of its sun. Did it suffer the same fate?*

lot of people who think there aren't going to be any atmospheres around hot rock planets." Although small, we have seen how red dwarfs have a propensity to fire off huge flares. In the visible spectrum that humans can see, M stars appear dim, which is why we cannot see all the red dwarf stars in our night sky. But they are far more energetic in other parts of the spectrum, to the detriment of nearby terrestrial planets. "With M stars you can be getting very little visible light, but X-rays are blasting away, driving off any atmosphere," Mandell says.

University of Colorado researcher Allison Youngblood adds, "M stars live a long time. They're in slow motion. Since everything takes so long, it seems like bad news for life. The nearby planet will continue to be bombarded by particles, and high levels of X-ray and ultraviolet radiation, which act to destroy atmosphere. Photons and particles heat the upper levels of atmosphere. When you heat something, it expands. As the atmosphere expands, it's no longer held on to as tightly by the planet's gravity well, and can be lost by many mechanisms. Planet heating causes atmospheres to puff up, and outer layers can be picked off." This action of "picking off" the air is called atmospheric erosion. The longevity of red dwarfs may also be good news for companion exoplanets, Youngblood says. "Planets are constantly replenishing their atmosphere through outgassing or delivery of volatiles [e.g., comets]. So it's an open question: Could the lost atmosphere be replenished over time? Once you get past the first 1 or 2 billion years, the star is still more active than the Sun, but then you have billions or trillions of years to replenish your atmosphere; you have a lot of time." Many factors affect the potential of atmospheric loss for exoplanets in orbit around M stars. In the case of Teegarden's worlds, a 2019 study put the chances of an atmosphere on Teegarden b at a scant 3 percent, while Teegarden c was given only a 2 percent chance. With this range of estimates, it is obvious that we have much to learn about the Teegarden family of planets.

So, in the search for life-friendly worlds, we have this conundrum: many rocky, Earth-size worlds orbit within their star's habitable zone, but their suns pour out searing flares and deadly X-rays. Planets in close orbits have the additional problem of tidal locking, so that one face is constantly cooked while the opposite face shivers in eternal darkness. But some planets have a built-in system for moderating climate: atmospheric circulation. On our own planet, tsunamis of air move across the face of the Earth, mixing warm air from the tropics with cold air from high latitudes. A similar process may be at work on exoplanets that have retained atmospheres, according to a new study from the Catholic University of Leuven in Belgium.

Researchers at the university ran 165 computer models of climate change, applying them to various possible

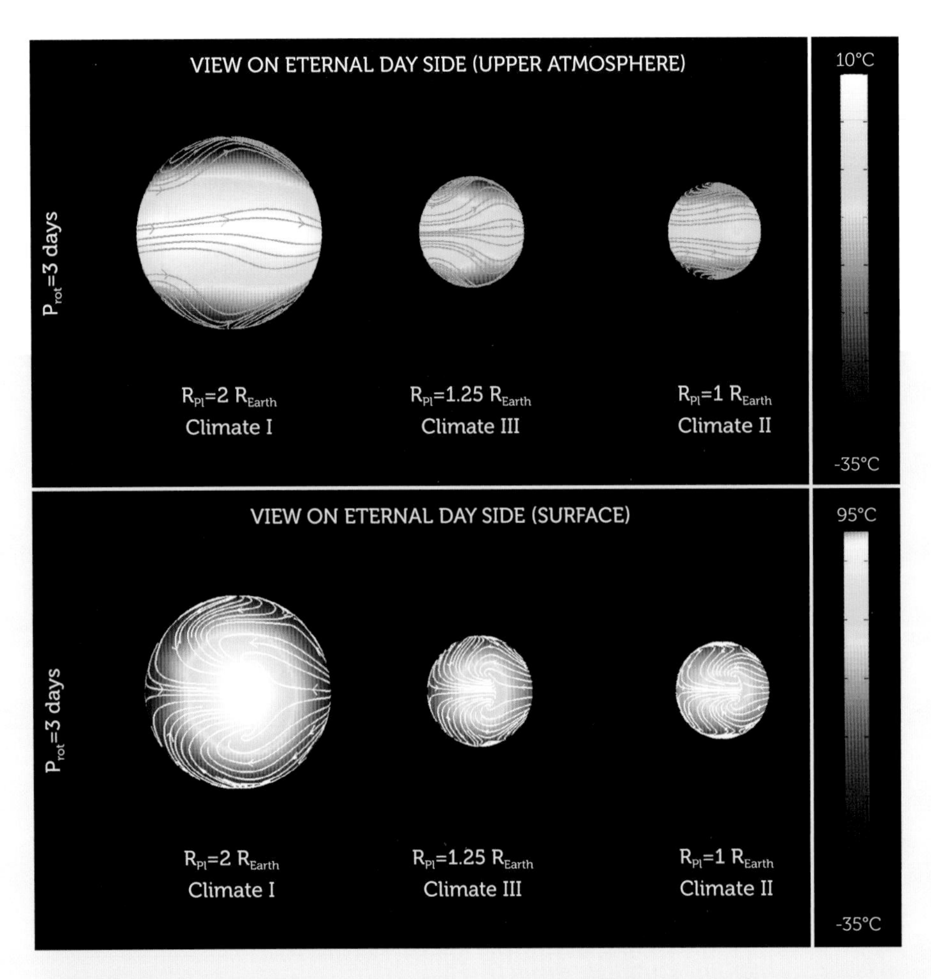

Left: *Researchers at the Catholic University of Leuven in Belgium studied three possible models of global wind systems on rocky exoplanets of different sizes orbiting red dwarfs. Climate I has a strong eastward wind jet along the equator. Climate II has two wind jets at higher latitudes. Climate III has wind jets at higher latitudes and a weak wind jet along the equator. Climate I disturbs the planetary "air conditioning," which cools the eternal dayside. Climates II and III don't disturb these important airflows, resulting in more habitable environments.*

Below: *Two familiar worlds have superrotation: Venus and Titan. In both cases, a great wave of air moves across the face of the globe in a matter of days. Similar currents may disrupt the climate on exoplanets.*

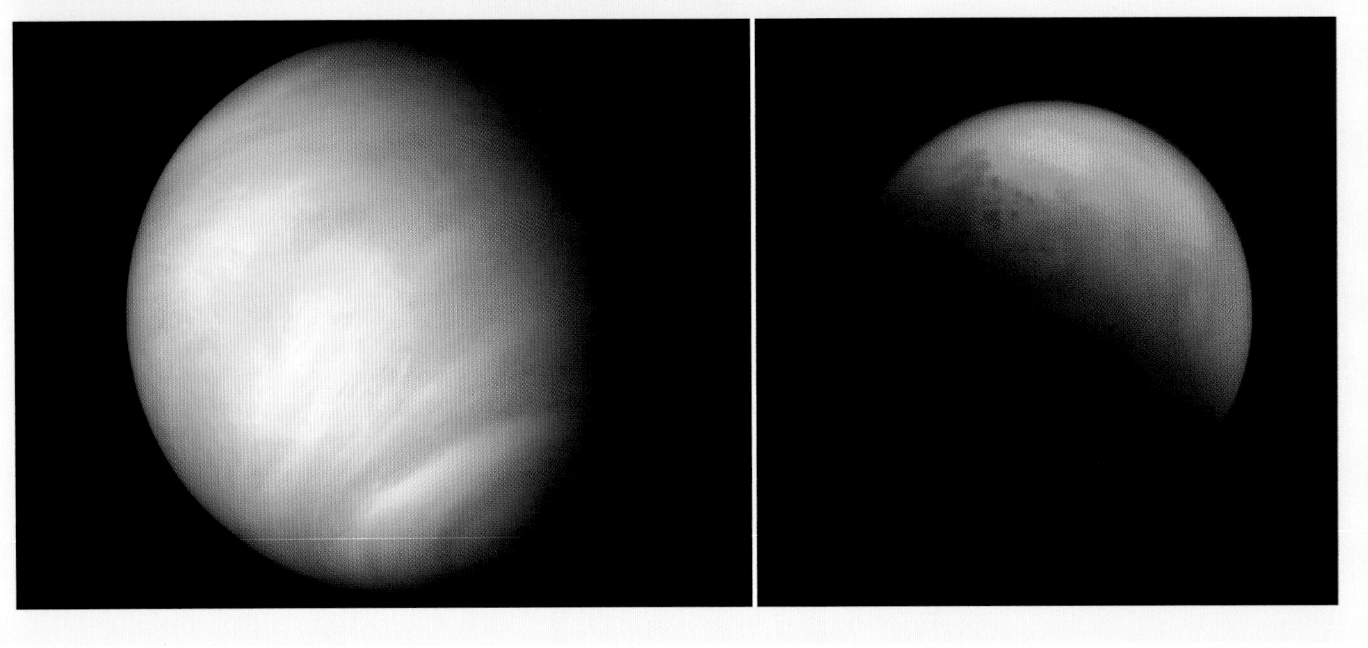

Above: *A host of potentially habitable planets parade across this chart by the Planetary Habitability Laboratory. Planets are arranged by distance from Earth, indicated within brackets in light-years. Each planet also notes the type of star they orbit (for example, the Earth is labeled "G—Warm Terran"). Teegarden b is at upper right, while its sibling Teegarden c is at the beginning of the second row. Remember that in many cases, a planet's characteristics are only approximate.*

exoplanets. Their study built 3D models of exoplanet atmospheres using different rotation periods (from one to 100 days) and planet sizes (from smaller than Earth to twice as large as Earth). The computer modelers found that rocky planets tended to have three potential climates; two of those climates could support life.

One scenario was fatal to life as we know it. In this case study, planets had orbits lasting fewer than 12 days. Computers showed the development of a strong eastward flow of air, called "superrotation," in the upper atmosphere. The wind jet disrupted the global circulation of air, resulting in dayside temperatures too hot for life. The researchers did find life-sustaining climates in two other scenarios. One model developed two weaker wind jets closer to the poles, while the second combined a weak superrotation stream with high-latitude flows. These configurations of wind currents did not disrupt the global atmospheric mixing, so the hypothetical planets remained livable.

The study of exoplanets is dynamic. Totals continue to change, but studies up to late 2019 brought the running count of planets with high enough ESI ratings to about 21. According to the University of Puerto Rico, the count of planets with slightly lower ESI numbers (less likely to be rocky but within limits of habitable zones) adds another 34 to the list.

At right is the current tally of exoplanets with possible Earthlike environments. The list includes such concepts as distance from the planet's sun, mass, and star type. The second set of worlds are Super Earths, gas giants or other non-terran planets that orbit within habitable zones. Many Super Earths are large enough to maintain conditions quite different from Earth's benign environment, and may be closer in nature to a warm version of Neptune or Jupiter. Notice that planets orbiting Sun-like G stars are rarely found on the list because it takes years to see repeat performances of their years long transits.

Below: *The star Gliese 667C hovers in the cloudy sky of its exoplanet Gliese 667C c, a planet within the habitable zone. It is likely that this exoplanet is a rocky terran world, and experts have given it an ESI of at least 0.71. The two other stars in the system, Gliese 667B and A, are seen to the upper right.*

EXOPLANETS WITH PROBABLE ROCKY COMPOSITION, MASS LESS THAN 3× EARTH'S, AND POSSIBLE EARTHLIKE CONDITIONS

For reference, Earth:	G star	warm terran	orbit period (days) 365	ESI 1
1. Teegarden b	M star	warm terran	orbit period (days) 4.9	ESI 0.93
2. K2-72e	M star	warm terran	orbit period (days) 24.2	ESI 0.90
3. Gliese 3323 b	M star	warm terran	orbit period (days) 5.4	ESI 0.90
4. TRAPPIST-1 d	M star	warm sub-terran	orbit period (days) 4.0	ESI 0.89
5. Gliese 1061 c	M star	warm terran	orbit period (days) 6.7	ESI 0.88
6. TRAPPIST-1 e	M star	warm terran	orbit period (days) 6.1	ESI 0.87
7. Gliese 667Cf	M star	warm terran	orbit period (days) 39	ESI 0.87
8. Proxima Centauri b	M star	warm terran	orbit period (days) 11.2	ESI 0.87
9. Kepler-442 b	K star	warm terran	orbit period (days) 112.3	ESI 0.85
10. Gliese 273 b	M star	warm terran	orbit period (days) 18.6	ESI 0.84
11. Gliese 1061 d	M star	warm terran	orbit period (days) 13.8	ESI 0.80
12. Wolf 1061 c	M star	warm terran	orbit period (days) 17.9	ESI 0.79
13. Gliese 667C c	M star	warm terran	orbit period (days) 28.1	ESI 0.78
14. Tau Ceti e	G star	warm terran	orbit period (days) 162.9	ESI 0.74
15. Kepler-1229 b	M star	warm terran	orbit period (days) 86.8	ESI 0.73
16. Gliese 667C e	M star	warm terran	orbit period (days) 62.2	ESI 0.71
17. TRAPPIST-1 f	M star	warm terran	orbit period (days) 9.2	ESI 0.70
18. Teegarden c	M star	warm terran	orbit period (days) 11.4	ESI 0.69
19. Kepler-62 f	K star	warm terran	orbit period (days)267.3	ESI 0.69
20. TRAPPIST-1 g	M star	warm terran	orbit period (days)12.4	ESI 0.59
21. Kepler-186 f	M star	warm terran	orbit period (days) 129.9	ESI 0.58

Top 20 exoplanets greater than 2.5× Earth's diameter with masses greater than 5× Earth's, unlikely to be rocky. Many are probably gaseous water worlds. Notice the pattern of exoplanets at warmer stars (i.e., K and G) with longer orbits in habitable zones.

1. Kepler-452 b	G star	warm superterran	orbit period (days) 384.8	ESI 0.83
2. Kepler-62 e	K star	warm superterran	orbit period (days) 122.4	ESI 0.82
3. Kepler-1652 b	M star	warm superterran	orbit period (days) 38.1	ESI 0.82
4. Kepler-1544 b	K star	warm superterran	orbit period (days) 168.8	ESI 0.80
5. K2-3 d	M star	warm superterran	orbit period (days) 44.6	ESI 0.80
6. Kepler-296 e	M star	warm superterran	orbit period (days) 34.1	ESI 0.80
7. Kepler-283 c	K star	warm superterran	orbit period (days) 92.7	ESI 0.79
8. Kepler-1410 b	K star	warm superterran	orbit period (days) 60.9	ESI 0.78
9. Kepler-1638 b	G star	warm superterran	orbit period (days) 259.3	ESI 0.76
10. K2-296 b	M star	warm superterran	orbit period (days) 28.2	ESI 0.76
11. Kepler-296 f	M star	warm superterran	orbit period (days) 63.3	ESI 0.75
12. Kepler-705 b	M star	warm superterran	orbit period (days) 56.1	ESI 0.74
13. Kepler-440 b	K star	warm superterran	orbit period (days) 101.1	ESI 0.74
14. Kepler-1653 b	K star	warm superterran	orbit period (days) 140.3	ESI 0.74
15. Gliese 832 c	M star	warm superterran	orbit period (days) 35.7	ESI 0.73
16. Kepler-1606 b	G star	warm superterran	orbit period (days) 196.4	ESI 0.73
17. Kepler-1090 b	G star	warm superterran	orbit period (days) 198.7	ESI 0.72
18. Kepler-61 b	K star	warm superterran	orbit period (days) 59.9	ESI 0.72
19. Kepler-443 b	K star	warm superterran	orbit period (days) 177.7	ESI 0.71
20. K2-18 b	M star	warm superterran	orbit period (days) 32.9	ESI 0.71

EYEBALLS: BIZARRE PLANETS WITH 20/20 VISION

The tidal locking so common to planets in close orbit to their red dwarf stars results in conditions not found in our solar system's planets. The Earth's Moon is tidally locked, so that we see only one face from our standpoint, but the Earth is not locked to the Moon, which turns once each month, compared to Earth's 24-hour spin. The closest true planet we have to these synchronous conditions in our own system is Venus, whose lazy 116.6 Earth-day spin is just a little shorter than its 225-day year. The situation means that Venus actually turns slowly backward (retrograde). Pluto and its comparatively huge moon Charon provide a closer parallel. The two bodies circle around each other, with the same face turned in toward the other throughout their orbit. The strange dance of tidal locking results in conditions perhaps best seen on the theoretical "eyeball" planets.

As we saw with Teegarden b and c, a planet that is tidally locked to its star will witness some wondrous things in the sky. From the daylit side, the sun appears locked in place in the sky, never moving or setting. A viewer on the nightside would face eternal darkness—unless an alien moon accompanies the planet, casting its feeble moonlight onto a landscape of cold desolation. At the terminator, that twilight between full day and full night, the star would hover at the horizon, never rising or setting, but perhaps warming the region into a narrow, temperate zone.

Right: *Eternal twilight visits a ring around tidally locked planets of M stars. This terminator region, between full day and full night, will see spectacular views of the nearby star. Depending on the planet's distance to its sun, the surface may see conditions quite Earthlike in temperature. This in-between territory will have temperatures averaged between the more extreme day and night hemispheres.*

Left: *Because the Earth's Moon is tidally locked to its parent planet, the Earth appears to remain stationary in the sky. While Earth would go through phases like the Moon does in Earth's sky, from gibbous to full to crescent, it would stay in the same spot from day to day, year to year.*

. . . A PLANET THAT IS TIDALLY LOCKED TO ITS STAR WILL WITNESS SOME WONDROUS THINGS IN THE SKY.

But a locked world will have more curiosities on its surface. Smaller worlds will be rocky. On the side facing the sun, even if the planet orbits within the habitable zone, desert conditions will likely occur in which the sun continually blazes overhead. But closer to the nightside, the environment will become more clement, perhaps allowing for the flow of liquid water within that twilight zone between dark and light. The nightside is another case: if the planet has enough atmosphere, and if that atmosphere has strong enough wind currents, the nightside may remain at temperatures benign enough for life. But if the air is thin or stagnant, temperatures will drop to well below freezing, even if the planet is on the inner edge of its star's habitable zone.

A second example of an eyeball planet is a larger terrestrial close in size to a Super Earth. Planets like these will be water worlds. The night hemisphere may well suffer arctic conditions, with oceans frozen solid, but on the dayside, the oceans will thaw into life-friendly seas. At the region pointing directly at the sun, clouds will likely burn away and water temperatures may rise close to boiling point.

In our hunt for life-hosting planets, size does matter. Large planets tend to lack a solid surface, and the smaller ones lack enough gravity to retain a substantial atmosphere. This makes the search for an Earth-size exoplanet all the more important. The Earth-sized planets that

Left: *Large enough to be a water world, some Super Earths may orbit at the outside edge of their star's habitable zone. Seas are likely frozen on the permanently shadowed nightside, but open ocean may exist on the hemisphere facing the sun. Because these worlds are in close orbits around red dwarf stars, they are tidally locked.*

Above: *Oceans—most of them frozen—congregate on the nightside of a rocky terrestrial in a habitable zone, but surface temperatures on its dayside become progressively warmer in areas exposed to the sun's direct light. A harsh desert faces the nearby star at the center of the planet's illuminated hemisphere. Along the terminator, a "ring of life" may form in the region permanently straddling the cold night and hot, desiccated day hemispheres.*

have been found were, for a time, all parts of planetary systems around red dwarf stars.

In December 2011, the Kepler Observatory team announced the discovery of the first Earth-size planet orbiting a Sun-like star called Kepler-20. Kepler-20's family of planets consists of at least five worlds, each one closer than Mercury is to the Sun. Two of the planets, Kepler-20 e and f, are about the size of Earth. The other three planets are too large to be terran in nature, and are probably gaseous planets the size of Neptune. They are Kepler-20 b, c, and d. The planets are arranged, from closest to the star to farthest, as b, e, c, f, and d, with small and large planets alternating.

Finding the first Earth-size planets at a star similar to the Sun was an exciting moment, but the planets are not similar to Earth at all. Kepler-20 e's path keeps it within 8.9 million km (5.5 million miles) of the star, giving it surface temperatures of 750°C (1,380°F). Kepler-20 f is almost identical to Earth in size, but not much more temperate than its sibling. At a distance of only 21 million km (13 million miles), surface temperatures on this "Earth twin" seethe at 425°C (800°F). Rocks on both terran worlds would glow red-hot, although the planets are probably tidally locked and may be cooler on the nightside. Still, none of Kepler-20's Earth-size planets are in the habitable zone of their Sun-like star. The distant twin Earths are undoubtedly desolate, desiccated places. But there are Earth-size planets within habitable zones, and two of them orbit a star called TRAPPIST-1.

Above: *The Earth-size Kepler-20 e and f compared to Venus and Earth. Kepler-20 e is so close to its star that it glows red-hot, while Kepler-20 f, nearly identical to Earth in size, may have a hot, dense atmosphere more similar to that of Venus than Earth.*

AN ETERNAL NIGHT SKY

Night on the tidally locked Kepler-20 e would sometimes offer a spectacular sight. Although the hemisphere facing away from its sun never sees daylight, several gaseous giant worlds populate the night sky, each about the size of Neptune, orbiting close by. Kepler-20 c, with the nearest orbit outside that of Kepler-20 e, passes within 5 million km (3 million miles) of the planet. It would appear nearly twice the size of a full moon in Earth's sky. A second Neptune-class world, Kepler-20 d, orbits farther away, and the Earth-size Kepler-20 f passes the field of view, dwarfed by the other two. Another giant world, Kepler-20 g, does not orbit in the same plane and would rarely make its presence known.

Above: *One side of Kepler-20 e never sees its M star: the Earth-size planet is tidally locked, so one face remains continually turned away from the light of its sun. Looking out into the night sky, Neptune-size Kepler-20 c and d would cross the sky occasionally, as would the second Earth-size planet, Kepler-20 f. At times, the three planets would undergo a conjunction, an event in which they appear to congregate near each other.*

VACATION SPOTS AMONG THE RED DWARF PLANETS

Burning like a cool ember some 39 light-years away, the M star TRAPPIST-1 lies in the constellation of Aquarius. The diminutive star is just one-twelfth the Sun's mass and is only slightly larger than Jupiter. Thomas Fauchez studies the TRAPPIST-1 system at NASA Goddard, and it intrigues him. "TRAPPIST-1 is a very ancient star. Its planets may be 6–9 billion years old, twice as old as our own worlds."

The humble TRAPPIST-1 has something no other solar system has: a host of Earth-size planets. The planetary menagerie was first discovered by the TRAnsiting Planets and PlanetesImals Small Telescope (TRAPPIST) and observations were expanded by the Spitzer Space Telescope, Very Large Telescope (Cerro Paranal, Chile), UKIRT (Hawai'i), Liverpool Telescope, and William Herschel Telescope (both in the Canary Islands). In our search for an Earth 2.0, TRAPPIST-1 offers many candidates. "TRAPPIST-1 d is somewhere between Venus and Earth for surface heat," Fauchez explains. "1 c is like Venus. If we could, by magic, move Earth to its place, the whole ocean would evaporate. 1 b is even hotter, about 126°C [259°F]." Its surface is not hot enough for a magma ocean just from its star's heat, but like its siblings, the planet has strong tidal heating, not only from the nearby red dwarf but also from its planetary companions. Considering the added heat from these gravitational interactions, the surface may be melted into a magma ocean.

All is not fire and brimstone at TRAPPIST-1. Planet 1 e has temperatures similar to Earth, Fauchez says. "1 e can have liquid water on the surface. It's 30°C [86°F] at the warmest. 1 f is between −50°C and −23°C [−58°F to −9°F], and 1 g is even colder."

The TRAPPIST-1 system's remarkable seven planets, all Earthlike in size, circle in tight, fast orbits around their dim sun, all of them closer than Mercury is to our Sun. The closest skims just 750,000km (466,000 miles) away. Surface temperatures on the farthest member of the TRAPPIST-1 family, a scant 9.27 million km (5.76 million miles) distant, hover at roughly −104°C (−155°F). The outer world is just one-tenth the distance that Earth is from the Sun, but TRAPPIST-1 is too weak to warm it beyond the small world's frigid temperatures. Both planets d and e may orbit within the habitable zone. Surface temperatures on planet d are estimated to reach a few degrees above freezing, and planet e, while averaging as cold as −27°C (−17°F), is large enough for a substantial atmosphere that may pump up the surface heat through greenhouse effects.

The worlds of TRAPPIST-1 range in diameter from planet h's 9,810km (6,096 miles) to the 14,014km (8,708-mile) planet b. This compares to the Earth's diameter of 12,742km (7,918 miles). As is common to planets in tight orbits around stars, members of the TRAPPIST-1 family are tidally locked, keeping the same face pointed toward their sun. The system formed in the same way that our own

Right: *The Earth compared to the red dwarf star TRAPPIST-1 below, and our own Sun, above. The much-cooler TRAPPIST-1 is more dim and red than our G-type Sun, spanning just larger than the diameter of Jupiter.*

THE HUMBLE TRAPPIST-1 HAS SOMETHING NO OTHER SOLAR SYSTEM HAS: A HOST OF EARTH-SIZED PLANETS.

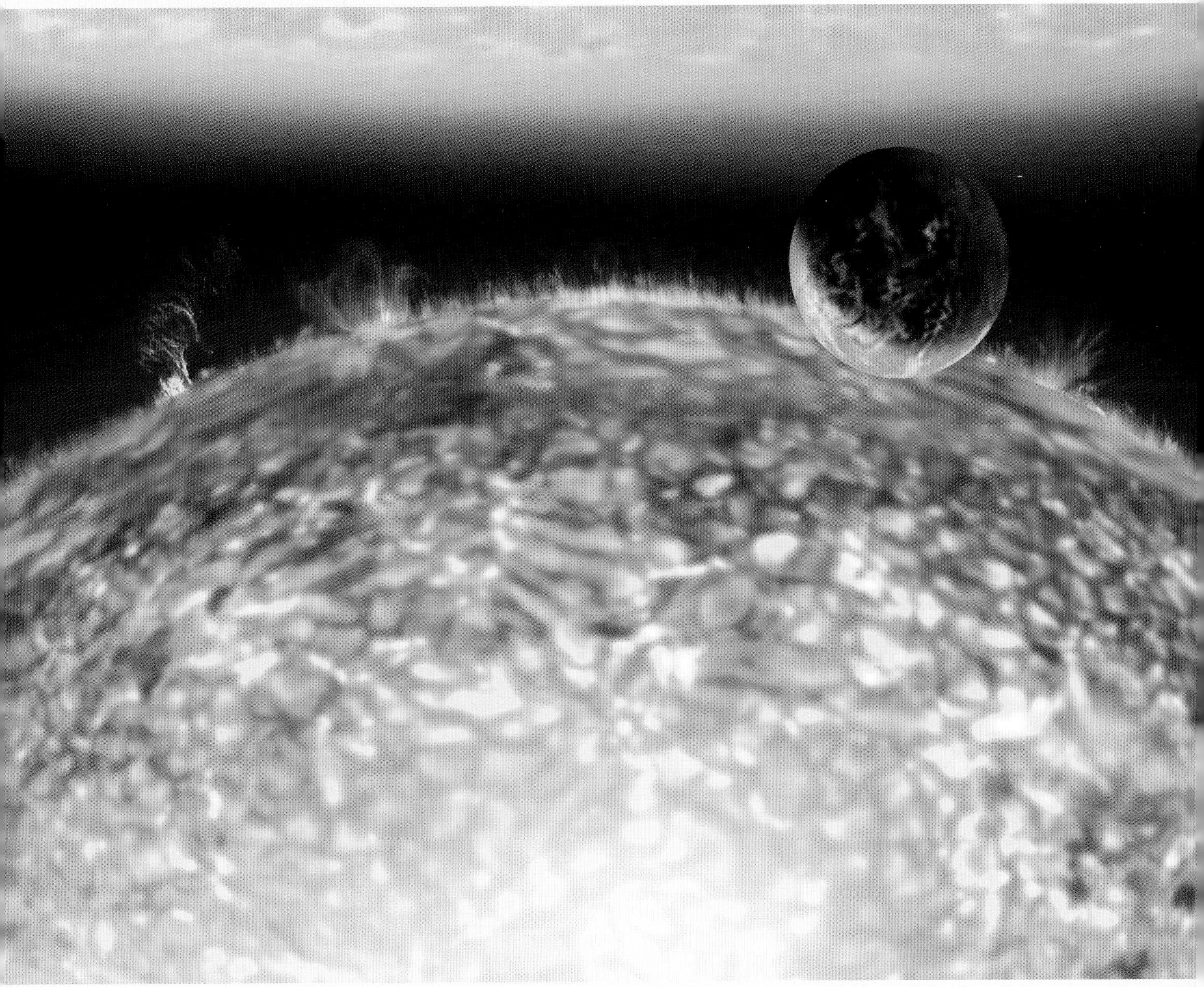

did, emerging from a disk of dust and gas. The Earth may not be far behind. The TRAPPIST-1 system shares a similar story, says Fauchez. "Simulations show that 500 million years is enough to put the whole system in synchronous rotation [with all planets facing the star]. We estimate that the age of TRAPPIST-1 is between 6 and 9 billion years old." Our solar system is roughly 4.6 billion years old. This also means that when the M star was very young and extremely active, these planets were already locked. Young

Left: *With an extended atmosphere denser than Venus's, TRAPPIST-1 b smolders beneath a searing sky, its surface simmering at nearly 2,000°C (3,600°F). TRAPPIST-1 b is probably large enough to retain its core heat, with active volcanism across the globe. In this view, we see white-hot rock glowing within surface fractures. The surface is too hot to support anything like basalt or granite mountains, but banks of lava may cool and thrust up here and there.*

Right: *Seen from an altitude of 800km (500 miles), the second planet of the TRAPPIST-1 system may be shrouded in Venus-like clouds. But other models suggest the possibility of a clear atmosphere of water vapor. This version of the planet envisions conditions in between. TRAPPIST-1 b glows in the near distance. Though TRAPPIST-1 c's surface bakes at 62°C (144°F), it may hold liquid water on its surface. Beyond orbits the superheated terrestrial world TRAPPIST-1 b against the backdrop of the Milky Way.*

stars go through an adolescent phase of brightening and extreme activity, spewing flares and radiation before settling down into stable adulthood. Fauchez points out that "For M dwarfs, the T-tauri phase can last for billions of years, so that means that for habitability, its complicated."

While some research predicts a thin atmosphere like that of Mars for TRAPPIST-1 d, Fauchez's models point to a world with a denser atmosphere, perhaps almost as dense as that found on Venus. If there are breaks in the clouds, views of nearby planets would

Above: *The third Earth-size planet from its star, TRAPPIST-1 d likely lies just within the habitable zone. Its size falls between that of Earth and Mars, so it may have a cool, thin atmosphere, with conditions similar to the warmer, wetter Mars of the ancient past. But its atmosphere may be far denser.*

Above right: *A pond reflects the icy pinnacles of a mountainous region on TRAPPIST-1 e. Nearby Earth-size planets float in an ice-fog sky. The closest, planet f, measures 1,609km (1,000 miles) farther across than Earth. Beyond it lie planet g (to the left) and the distant planet h (at right).*

be impressive. From planet d, planet c would appear nearly twice the size of a full moon in Earth's sky, affording a good view of its surface details.

Near the outer limits of TRAPPIST-1's habitable zone, planet e's surface may be warm enough, at least at lower altitudes, to support liquid water. In the high country, things are undoubtedly colder. For his money, Thomas Fauchez thinks this is the most Earthlike planet of the lot. "TRAPPIST-1 e is probably the only one with liquid water on the surface. It is also the right size/mass ratio

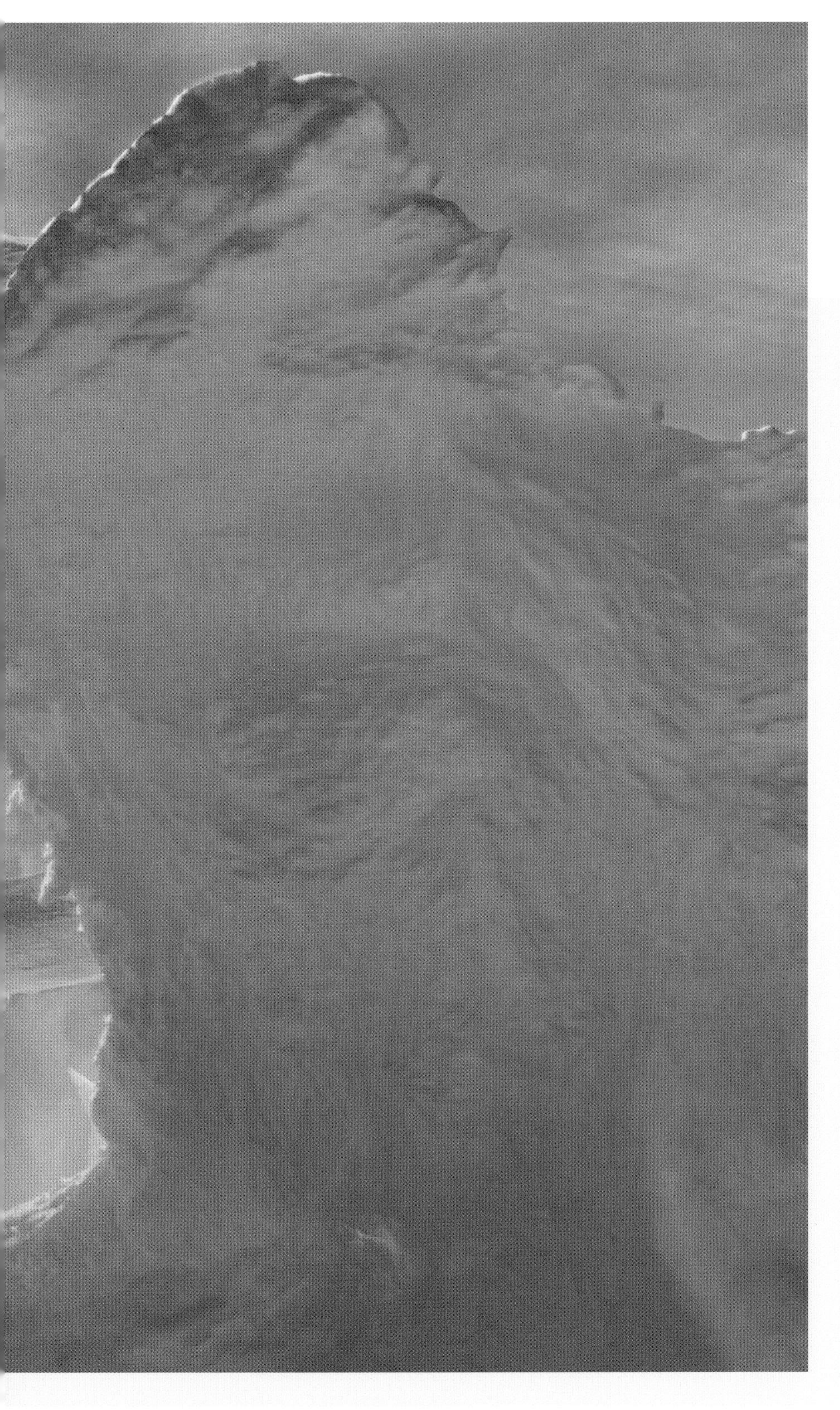

to have an iron core similar to Earth." Researchers are studying a spectrum of possible conditions on the next planet out, TRAPPIST-1 f. Without an atmosphere, the tidally locked planet's temperatures would hover around –59°C (–73°F), even on the daylit side. Yet because the planet is roughly 20 percent water, researchers suggest that TRAPPIST-1 f is blanketed in a dense atmosphere of water vapor that could pump temperatures to as high as 1,130°C (2,070°F). But the dense air is not all bad news: wind currents may bring enough warmth into the nightside, always turned away from the star, to create habitable environments in some locations.

Surface temperatures on the remotest TRAPPIST-1 world, planet h, may be similar to those in Antarctica during winter. With an atmosphere that may be denser than that of Mars, its icy surface may erode into exotic forms. From TRAPPIST-1 h, the entire planetary system of TRAPPIST-1 can be seen. Planet h circles its host star in just under 19 days. Like the ice moons of our own system, planet h might have a subsurface ocean,

Left: *Although larger than Earth, TRAPPIST-1 g is only 76 percent as dense. Estimates for surface conditions range from a thick atmosphere above an ice landscape to a frozen water world. This image of TRAPPIST-1 g depicts the latter. In the sky, TRAPPIST-1 f blocks out some of the star's face in eclipse, as does the more distant TRAPPIST-1 e. In the TRAPPIST-1 planetary system, eclipses are common.*

Left: *A dense blanket of air warms the surface of Earth-sized TRAPPIST-1 f in this rendering of one possible environment of the fifth planet from the red dwarf. Planet f's orbit is near the outside edge of the habitable zone of its sun, so liquid water is possible on the surface. If TRAPPIST-1 f is cocooned in atmosphere more dense than that of Earth, landscapes may include abundant water in rivers, lakes and even seas. Just how much liquid water exists on the surface is open to interpretation, awaiting more data from probes like the TESS observatory.*

especially if tidal forces from nearby planets are heating its interior.

But what about that problem of atmospheric erosion? Researchers like Mandell and Youngblood suspect planets like those in the TRAPPIST-1 system will have been stripped of their atmospheres. Thomas Fauchez is not so sure; the worlds of TRAPPIST-1 have masses that indicate the presence of water, and that may be a game changer. "We may find a lot of airless planets, but I still think in some configuration they can preserve some atmosphere. The problem is that TRAPPIST-1 is an active star, so we have two different sides that could have strong positions. TRAPPIST-1 is active, so it can have a lot of flares, which is bad, but in the meantime the planets may have a lot of water, which is good for life, so you have a huge fight." That fight promises to continue until we gain more insights into the true nature of this unusual planetary system.

Below: *Seen from the surface of TRAPPIST-1 h, the outermost world in the TRAPPIST-1 system, planet g eclipses the dwarf sun TRAPPIST-1. Three quarters the diameter of Earth, TRAPPIST-1 h is the smallest of the system's planets. It is also less dense, and is probably covered with a thick ice shell.*

Right: *A dinner table full of stars. M stars compared to the Sun (behind) are, from left to right: Barnard's Star, TRAPPIST-1, and Gliese 667C. All of the smaller stars are red dwarfs, considerably cooler and redder than the Sun.*

Below: *A view of the TRAPPIST-1 system from a vantage point in the neighborhood of TRAPPIST-1 f (at far right). Some models project that TRAPPIST-1 f is in the center of the habitable zone.*

BUT WHAT ABOUT MOONS?

Moons outnumber planets by more than 20 to 1 in our solar system. If that holds true for the rest of the galaxy, trillions of exomoons are out there—worlds where life could launch. Earthlike moons may well orbit Super Earths or gas giants in habitable zones, and those planets outnumber Earth-size habitable zone residents.

In our own solar system, the most Earthlike planets are Mars and Venus, but the moons of the outer solar system are prime targets in the search for life. Both Enceladus and Europa have deep, briny oceans in contact with life-enabling minerals in their rocky cores. The ices that seal the oceans in provide protection from radiation and the vacuum of space. At one time, small, cold moons seemed a fruitless place to search for life, but with the discovery of extremophiles—microbes living in extreme environments like the high-pressure environs of the ocean floor—the search for life has expanded to moons in the outer solar system (see Chapter 5).

Below: *Seen from the surface of an airless satellite, a sibling moon hosts Earthlike seas and weather patterns. Beyond it, another moon passes by, lifeless. Behind all of them lies a gaseous planet orbiting within its star's habitable zone, a location that enables life-friendly processes on its nearby moon.*

Right: *Just six light-years from Earth, the red dwarf Barnard's Star smolders in a dark sky. Orbiting the small star is a Super Earth 3.2 times the mass of Earth, circling outside of the star's habitable zone. Barnard's Star b is likely a frigid world with temperatures dipping to −170°C (−274°F).*

EXTRATERRESTRIAL LIFE ON OUR DOORSTEP?

THE CASE FOR PROXIMA B

In the summer of 2016, the world received exciting news. Researchers at the ESO confirmed a planet in the nearby Alpha Centauri system. The planet is known as Proxima b, and its discovery was the culmination of a 16-year study. Proxima b circles the M star Proxima Centauri, a cool red dwarf closer to our solar system than any other star—so close, in fact, that with some modest improvements in technology, we might be able to send a robot to its planetary companion in the span of less than 50 years. And the really great news is that the newly found planet may be Earthlike.

But just how Earthlike is this nearby world? What can we tell today, and what possible planetary types do the data truly represent?

THE MEASURE OF A NEW WORLD

The range of possible conditions on Proxima b is wild and woolly. Codiscoverer Michael Endl of the University of Texas's McDonald Observatory warns that, "Unfortunately, the data give us a pretty wide set of possibilities." Part of the reason that researchers cannot better constrain the planet's environment is that the world has not been seen directly, but rather by the technique radial—or Doppler—velocity technique (see Chapter 1). It's what Endl calls a "game of precision." As a planet orbits its parent star, its Doppler shift of light can be charted in much finer detail than a side-to-side visual movement seen through a telescope. As we've seen, the technique relies on a subtle change in starlight rather than a star's apparent movement against the background, which is much harder to resolve.

Doppler data indicate that Proxima b weighs in at a minimum of 1.3 Earth masses. Its 11-day, five-hour orbit places it within Proxima Centauri's habitable zone, where liquid water can exist. But the mass estimates are just that: estimates. Current data leave open the possibility that Proxima b is anything from a large, rocky terran world to a sub-Neptune behemoth, a bizarre hybrid planet somewhere between a supersize Earth and an ice giant gas world. The radial velocity method does not provide tight boundaries on the mass of the planet. Proxima b's mass could be as low as one

THE RANGE OF POSSIBLE CONDITIONS ON PROXIMA B IS WILD AND WOOLLY.

Below: *A photo of Alpha Centauri in visible and X-ray (inset) light. The two stars at center, Alpha Centauri A and B, appear as one in visible light (A). An X-ray view shows the separation of the two primary stars. Proxima Centauri is too dim to see in the large view, but is seen through the eyes of Hubble in the lower inset.*

Above: *If Proxima b has retained a substantial atmosphere, it may suffer conditions similar to those found on Venus. Venus is a twin of the Earth in size, but is at the inner edge of the Sun's habitable zone, and retaining enough atmosphere to trigger global greenhouse effects. The surface of Proxima b may be similarly heated. The planet's size assures that if it is rocky, it likely has active geology like volcanoes and geysers.*

Earth mass, but it could also be 100 or even 1,000 times that. Planets with larger masses are less likely as the orbit of a heavier planet would tend to be farther from its star than Proxima b is. Judging by other planetary systems, Proxima b is more likely a rocky world. From Kepler Observatory results, astronomers see more terrestrial-range planets orbiting M stars than those orbiting G stars like the Sun. It is easier to find terran worlds around red dwarfs because of their tight orbits and frequent transits. It takes many years to sense the repeated transits of an Earthlike world on a more distant course around a G star, but if the numbers are all taken into account, it still appears that red dwarfs hold the lion's share of Earth-size companions. It is not yet clear why, but the fact means that Proxima b is more likely in the neighborhood of a few Earth masses at most.

CELESTIAL WANDERINGS

Researchers do not yet have a measure of Proxima b's radius and they only have a range for its possible mass, two factors that can tell us a lot about a planet's nature. But if the planet is half again as massive as Earth, it may qualify as a sub-Neptune. If so, the planet may have migrated inward after forming farther away from its sun. The idea of planetary relocation falls into several theories, including such ideas as the Nice Theory and the Grand Tack (see Chapter 1). These revolutionary concepts came from an unlikely place: the study of the Kuiper Belt. Dynamic studies of the Kuiper Belt, a ring of icy bodies beginning at about the orbit of Neptune, intimate that our early planetary system had a different architecture than the one we see today. Some Kuiper Belt objects follow paths that are "in resonance" with Neptune. This means that for every three times that Neptune circles the Sun, a more distant object "in resonance" will orbit it exactly

twice. This is the case with Pluto, one of the Kuiper Belt objects in resonance with Neptune. Computer models reveal that the only way to trap Kuiper Belt objects in resonance with Neptune would be if Neptune started out closer to the Sun than it is today, and then slowly moved outward. Acting like a cosmic snow plow, Neptune would have spiraled out from a close orbit of the Sun, shoving a wave of Kuiper Belt objects ahead of it. Simulations indicate that Jupiter and Saturn may have migrated close to the Sun before ending up in their current orbits.

It is possible that Proxima Centauri's super-sized Earth may have started life father out in its planetary system, with a small hydrogen envelope—smaller than those of Neptune or Uranus. Early in Proxima Centauri's development, the primordial little world may have worked its way into close proximity to its sun. But unlike our gas and ice giants, Proxima b got trapped there, where it remains today.

MANY WORLDS TO CHOOSE FROM

Despite the uncertainties about Earth's nearest sibling, scientists are able to offer several possible scenarios for the planet's true nature. At the upper end of estimates, Proxima b may have no solid surface, resembling a warm version of an ice giant. Once a planet surpasses approximately 1.6 times the Earth's diameter, its nature will shift to that of a gaseous sub-Neptune. Researchers submit that planets remain rocky, with solid surfaces, until they reach a certain size, says Kepler scientist Elisa Quintana. "What people have figured out is that the transition from rocky Super Earth to gaseous sub-Neptune is about 1.6 Earth radius." If Proxima b measures a wider girth than that transitional size, it will likely not be Earthlike at all, but rather a gaseous world of wind and storm with a rocky, Earth-sized core. Still, even an alien sub-Neptune world may hold the promise of life: its more powerful gravity

Above: *Volcanoes on Earth play an important role in rejuvenating the environment and recycling the atmosphere. Some models suggest that terrestrials close to the scale of Earth may have active volcanism, while superterrans may have crusts too thick to generate plate tectonics and volcanic activity.*

Right: *The Earth's oceans stabilize our climate and even out temperature swings. Currents bring oxygen-making microbes, recharging the atmosphere. Water first condensed into surface seas on Earth over 3.8 billion years ago. Oceans on terran worlds larger than Earth may last even longer.*

Below: *If Proxima b is smaller than a Super Earth, but larger than our own planet, it may be a water world covered in a global ocean. Debates rage as to whether a planet-straddling sea would be a good place for life. Could life begin at the seafloor and develop forms that ultimately live near the surface? Are there enough minerals suspended in such an ocean that life could begin there? The Earth's oceans may well have been the cradle of life for our planet, making a water world a good target in our search for life in the galaxy.*

might gather one or more large moons into its realm, and these could harbor liquid water and Earthlike conditions on their surfaces.

NARROWING PROXIMA B'S NATURE

If the planet is smaller than 1.6 Earth radii, Proxima b's nature may be quite different from a sub-Neptune. A series of computer simulations carried out by France's Marseille Astrophysics Laboratory explored planetary compositions based on mass. Researchers studied various Proxima b radii ranging from 0.94 to 1.4 Earths. On the lower end of the scale, giving Proxima b a radius of 5,590km (3,473 miles), the planet would house a dense, metallic core comprising two-thirds of its overall mass, surrounded by a rocky mantle.

Intermediate sizes in the model indicated that some water would remain as liquid or break into hydrogen and oxygen atoms. On Earth, oxygen and water seem to be a winning combination for life. But oxygen is such a reactive molecule that it actually prevents the growth of many prebiotic molecules. Astrobiologists suggest that the presence of oxygen early in a planet's environment could short-circuit life's operations. Life on Earth appears to have formed in an oxygen-free environment. It could be that Proxima b has abundant water and oxygen but is not habitable.

A soggier scenario envisions Proxima b as a larger water world with a 1.3 Earth mass, washed in globe-encircling oceans hundreds of kilometres deep. The Marseille study suggests that with a radius 1.4 times that of Earth (approximately 8,920km / 5,543 miles), Proxima b would possess a rocky core making up 50 percent of its mass, blanketed by an ocean that makes up the other half. A Harvard study showed that oceans on rocky planets with two to four times Earth masses are even more stable than those on Earth, lasting at least 10 billion years (assuming the primary star does not swell to the red giant phase and boil the oceans away). The biggest case studied was a planet with five times the Earth's mass. That theoretical planet's thick crust staved off early volcanic activity. This would be bad news, as the Earth's own oceans are thought to have come at the hands of water vapor released through volcanism. But the Harvard model indicated that later eruptions on their theoretical world recharged its atmosphere, establishing stable oceans over the long term.

PROXIMA'S FAMILIAR PROBLEM

But is Proxima b habitable in an Earthly sense? The key may lie within its own sun, a star much cooler than our own. Proxima Centauri joins the majority club of cool red dwarfs scattered across the galaxy, but it has not always been so cool. Early in the evolution of a red dwarf, the star puts out a prodigious amount of energy before it calms down to its more familiar lukewarm state. This energetic brightening stage is called the T-tauri phase. If Proxima b has spent a lot of its life in the location where it is today, it suffered through torrential waves of heat and radiation in its formative years.

Proxima Centauri's frenetic adolescence could do several interesting things to a nearby planet. It could completely blow away all the water, leaving behind a bone-dry, desiccated planet not conducive to life. But what if Proxima b began as a much larger sub-Neptune? In Proxima Centauri's stellar youth, its radiation would strip the nearby planet of volatiles like water and lighter gases (helium, hydrogen), leaving an oceanic or rocky world. This transition could, under the right circumstances, transform a Neptune-like world into a habitable one.

Radiation is another matter. Proxima Centauri frequently tosses out energetic solar flares, says Kepler scientist Thomas Barclay. "Proxima Centauri is a high flare activity star. It is not hugely atypical, but it is on the higher side. Red dwarf stars like Proxima Centauri are tempestuous, even into their mature years. They can have very high-energy explosions on their surfaces. Proxima Centauri has massive flares several times a year. Once-in-a-century events on our Sun are once-every-six-month events there." These potentially deadly flares send waves of X-rays and ultraviolet radiation across the face of Proxima b. At times, the surface of the planet would sustain X-rays 400 times as strong as those the Earth receives from the Sun. "Not only do you have these frequent flares," Barclay says, "but you've got a planet really close by. In a star like Proxima's case, you have to be closer to the campfire to keep warm."

But high radiation levels are not a show-stopper for Proxima b's potential

ONCE-IN-A-CENTURY EVENTS ON OUR SUN ARE ONCE-EVERY-SIX-MONTH EVENTS THERE.

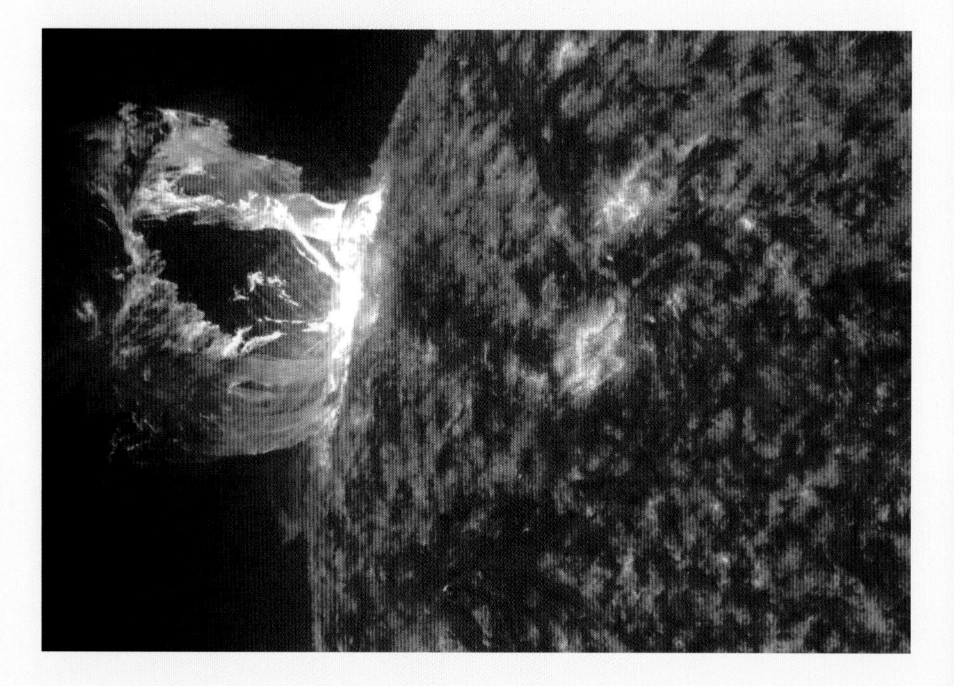

Below: *A dramatic solar flare on the Sun can be a commonplace occurrence on an M star. A medium-sized flare exploded on the Sun on April 16, 2012. This looping arc of particles, called a prominence, formed at the same time. Such prominences are often associated with flares.*

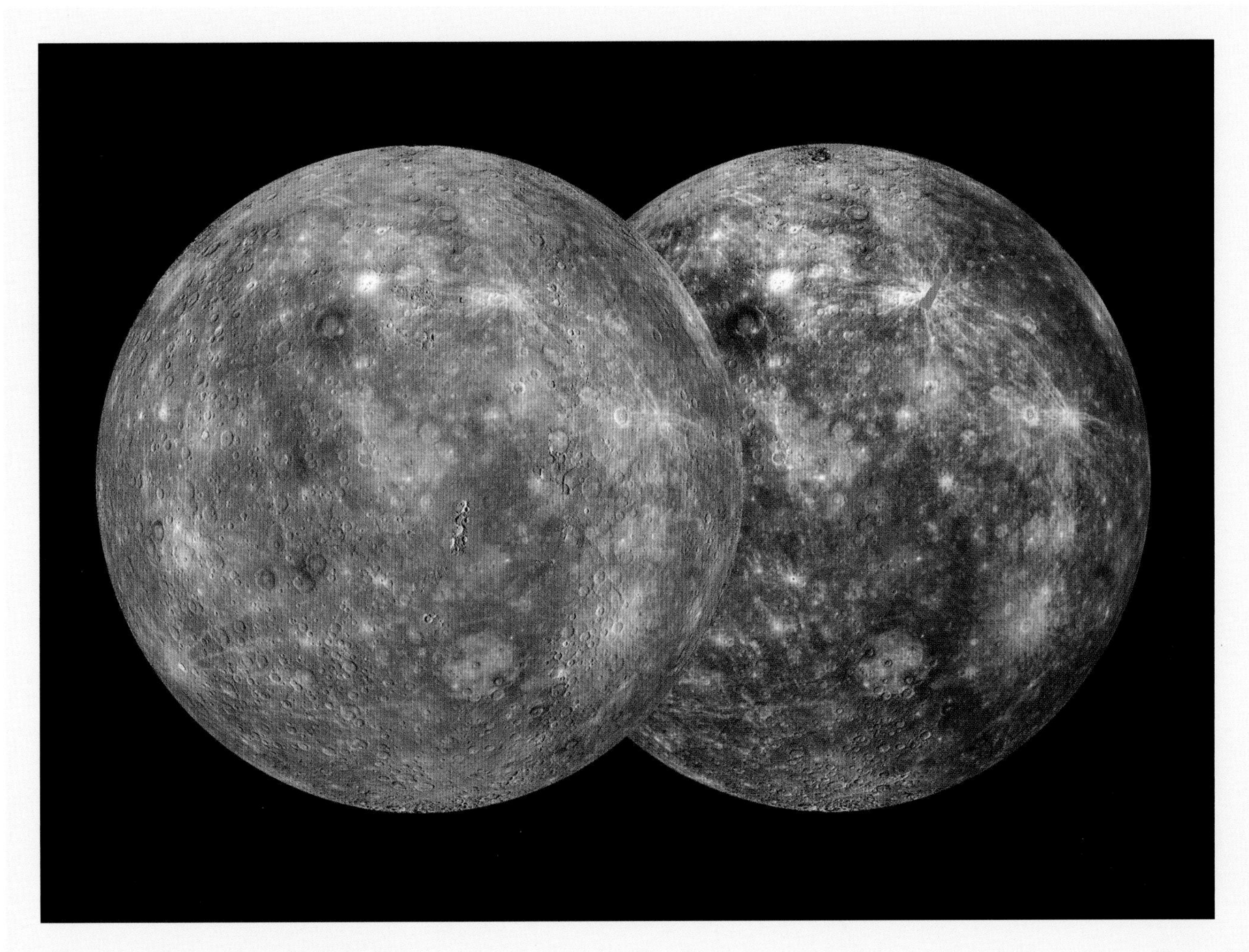

MERCURY IN ORBIT

Mercury is nearly tidally locked to our Sun, making three rotations for every two orbits. This results in some strange events in its sky. At some locations on Mercury, and at certain times in its year, the Sun appears to rise, then reverse itself and set in the place where it rose, before ascending again to make the trip to the other horizon. In time, the gravitational field of the Sun will force Mercury into a synchronous, Sun-facing orbit, as we see in the case of worlds in close orbits of red dwarfs. Venus, too, will succumb to the Sun's tidal forces, eventually becoming a world locked in its Sunward gaze, with eternal night on one hemisphere and blistering eternal day on the other. These normal (left) and false color (right) views were assembled from the MESSENGER spacecraft mission. The smooth, gray areas are regions where data are missing. While planets in close orbits around red dwarf suns tend to turn once for each orbit around their star, our solar system has no such worlds. The closest is Mercury, which turns in a lazy, 59-day rotation.

Left: *An ice crystal halo encircles the sun above an Earthlike Proxima b. Alpha Centauri A and B glow in the distance at upper right. Although Proxima Centauri sends out deadly flares, Proxima b may have natural barriers to the associated radiation, such as water vapor and a strong magnetic field. With just the right conditions and natural balances, Proxima b may be the nearest Earthlike planet to our own.*

for life. A super-sized Earth may have a super-core. If the planet hosts a large enough molten metallic core, a strong magnetosphere could shield Proxima b from much of its sun's radiation. Allison Youngblood explains, however, that a strong magnetic field may be a double-edged sword. "Magnetic fields can be helpful and harmful. What they do is to funnel particles." Flares are mostly photons—particles of light—but often accompanying them are coronal mass ejections, deadly "packets" of radioactive material from the star. Incoming particles are trapped inside the planet's magnetic field. "All these particles interact with the magnetic field. The particles are influenced by the field and are funneled to the poles. That is helpful, because you are localizing the damage toward the poles. That may be good. But the way that magnetic fields are not helpful is that they make the planet look bigger to these charged particles. When you have a magnetic field, your 'catcher's glove' is bigger, so instead of missing the planet, more particles are snagged. So that protective field is also grabbing more charged particles than would otherwise intersect the planet. It's an open question as to where this crossover point is between being helpful and being harmful."

If life did take hold on Earth's nearest sibling, what kind of world would it inhabit? Beneath its alien skies, any life on Proxima b could see a plethora of landscapes: frozen wastelands, rugged mountains, parched deserts, or oceans from horizon to horizon. What strange and wondrous forms would life take under the angry glare of a red dwarf sun?

Proxima Centauri itself would appear three times the diameter of the Sun in our sky. With an orbit so close to its parent star, Proxima b may be tidally locked, keeping the same face toward its sun. If

Above: *A terran Super Earth orbits a K star—a star slightly smaller and cooler than the Sun—in this imaginary view of what may be a commonplace vista across the galaxy. Will planets circling K-type stars have a better chance of hosting life than their cool cousins orbiting the M stars?*

this is the case, only one hemisphere of the globe would ever see the sun, while the other hemisphere would endure eternal night—but that long, dark night might not be so cold. Depending on its amount and circulation of atmosphere, this could result in a dry, hot hemisphere on one side and a chilled, eternally dark hemisphere on the other.

Planetary dynamicists suggest that Earth's big brother may take a less leisurely daily spin, perhaps akin to Mercury, which turns three times for every two revolutions around the Sun. If Proxima b moves in this way, with a day slightly shorter than its year, its atmosphere—if it has one—will circulate in patterns that could allow for quite temperate regions on the planet.

That atmosphere might also simmer as a pressurized hothouse, similar to what we see on Venus today. Proxima b also offers an important piece of the planetary puzzle in our search for distant Earths. It orbits the most common type of star in the galaxy and its probable size lies in the same range as the vast majority of exoplanets found so far, so Proxima b may be the poster child for Earthlike worlds in the universe. With its nearness to Earth and its possibilities for life-sheltering conditions, it may well be the best target for Earth's first interstellar probes. The timeline of these advanced missions may be shortened by Proxima b's very existence.

Among the exoplanets, Proxima b may not be the Earthlike world in which we seek life, but others are rising to the top of the list. Some of them reside in peculiar locations around even more peculiar suns. Our list of candidate life-bearing worlds so far is a short one, but it's growing. Along the way, researchers are uncovering places that are the stuff of science fiction . . . and nightmare.

4

EXOTIC EXOPLANETS AROUND STRANGER STARS

Proxima b may harbor conditions similar to Earth. Other worlds like TRAPPIST-1 e might share natures familiar to us. But these worlds—and their suns—are tame compared to bizarre planets orbiting some of the galaxy's strangest stars. In all corners of the Milky Way, we find titanic suns brimming with energies far greater than the sunshine we enjoy. Some stars squish into a football shape as they spin incredibly fast, creating an uneven illumination on their surfaces and across any nearby worlds. There are planets that wander too close to their powerful stars and still others that simmer with almost no heat at all, embers glowing in the dark skies of cold and dead worlds.

Left: *Seen from a nearby moon, Pi Mensae b glows with its own internal heat. The planet is so large that it borders on being a brown dwarf, a sub-star. Its eccentric orbit dooms the entire planetary system to a habitable zone devoid of planets.*

WORLDS TERRIFYING AND MYSTERIOUS

There is no morning on KELT-9 b. Like the Teegarden worlds, the planet is tidally locked, always facing its star. Its sun hangs low above the foggy horizon, glowing a brilliant purple, never moving from its spot. It looks odd, not completely round. The center of it is markedly darker than the outside, but this will change as the day progresses.

KELT-9 b is the hottest gas giant known, with daytime temperatures hotter than some M stars. Nearly three times the mass of Jupiter, its hydrogen atmosphere is streaming away from the planet at high rates, bringing with it vaporized iron and titanium. A comet-like tail of atmosphere trails from the cooked world as it races around its star once each 36 hours. And while KELT-9 b is an exotic world, its sun is even more bizarre.

KELT-9 is an A-type star, blazing at temperatures of 9,900°C (17,900°F), nearly twice that of the Sun. But temperature is just the beginning of the differences between the two. KELT-9 spins so quickly that it is egg-shaped, flattened significantly at the poles. NASA Goddard's Johnathon Ahlers studies A and F stars, some of the largest suns in the galaxy. These stars, Ahler says, are fundamentally different from smaller stars. "High-mass stars spin superfast. All stars start off spinning fast, but the lower-mass ones—like our Sun—slow down; the high-mass ones spin fast forever. It has a dramatic effect on these stars. They are oblate. They are gravity darkened, meaning that the poles of the stars are brighter."

This difference in brightness across the star also changes the heat falling on any nearby planets, Ahlers finds. "Many planets circumnavigating these stars are in highly inclined, even polar orbits." Because they follow these pathways, they circle over the bright poles twice each orbit, and then pass over the darkened equator twice per orbit as well. "Throughout the orbit, the planet varies in exposure to the star's dimmer equator and hotter poles."

This bizarre cycle—from the viewpoint of an orbiting planet—seems to be the norm for worlds around higher-mass stars. It's not an exotic scenario: nightmarish exoplanets orbiting similar stars include MASCARA 4b, a Hot Jupiter orbiting an A star in a 2.8-day cycle, as well as the gigantic Hot Jupiter Kepler 13, which orbits an A-type sun in a triple-star system.

Some planets may lie in orbits farther out from these stellar goliaths. For these stars, the habitable zone is football-shaped because it is pushed farther away at the warmer poles. Planets, too, can circle stars in oblong orbits, so some Earth-size worlds could witness the same exotic daily cycles as does KELT-9 b, but from a safe distance.

Right: *The brutal sunlight of KELT-9 shines down on a sweltering cloudscape of the gaseous colossus KELT-9 b. The exoplanet suffers such extreme temperatures and strong solar winds from its sun that it is losing atmosphere at a prodigious rate. The star's equator is gravity-darkened, a phenomenon that will affect the days and seasons of the broiling planet.*

These worlds would experience seasons quite different from the four we experience on Earth, Ahlers says. "The frequency of going from hot to cold happens twice in an orbit instead of once." In some scenarios, the close of spring would bring summer, followed by another spring; but the summer—that part of the cycle when the planet's summer hemisphere is pointed most directly toward its sun—would actually be cooler than either spring because it occurs over the darkest part of the star. "When it's trying to be summer, it's near the dimmer equator." Ahlers points out that this is only one of many possible scenarios. Gravity-darkened seasons could cause superheated summers; long, cold winters; or the "two-summer" effect.

With their odd, football-shaped Goldilocks zone and their extreme temperature changes, nearby planets might not be able to hold on to an atmosphere at all. The problem lies not only with the powerful force of the stellar winds flowing from these giant stars, but also from the kind of light they emit. Ahlers explains that, "Something that's very important to atmospheric scientists is that from this hot to cold gradient across the stellar surface, the amount of light changes the most in the ultraviolet. That's a very big deal for atmospheric chemistry. Ultraviolet light generates the ozone layer. It affects where the habitable zone may be. X-ray and ultraviolet light are ultimately what decides whether an atmosphere gets stripped or not. Having an ultraviolet radiance on a planet that doubles throughout the orbit—that's a huge deal and it's not clear at all what it would mean for these planets."

Still, Ahlers holds out hope that we will find harbors for life even among the A and F stars. While we have not yet, he suggests it's just a matter of time. "The path forward will lead us toward higher and higher mass stars over time. Our ability to find habitable zone planets will expand farther and farther out. We're looking at M dwarfs because we can find planets in short orbits in the habitable zone. But we want to look more at K and G stars—that's where the Sun is—we just need more time to get there." The habitable zone for A and F stars is out at a three- or four-year orbit, so researchers are not finding a lot of them yet. But Ahlers suggests that within a decade or so, people will be finding planets in habitable zones of gravity-darkened stars.

Above: *KELT-9 b sees varying views of its star. The star's rapid spin causes the oval-shaped KELT-9 to glow brighter at its poles than at its equator, a phenomenon called gravity darkening (polar view at right). Temperatures on KELT-9 b are high enough to vaporize materials in the giant planet's rocky core. KELT-9 b is one of the largest exoplanets known, large enough to border on being a brown dwarf (a sphere with a size and mass between planets and stars).*

While Kelt-9 b is an exotic world, its sun is even more bizarre.

Right: *The habitable zones of a Sun-like G star (above) and a gravity-darkened star, compared. Because its poles are hotter than its equator, the habitable zone of a gravity-darkened star is oblong. The irregular shape of the zone might make for interesting dynamics on a life-bearing planet, which would either follow the same egg-shaped orbit or experience wild differences in seasons.*

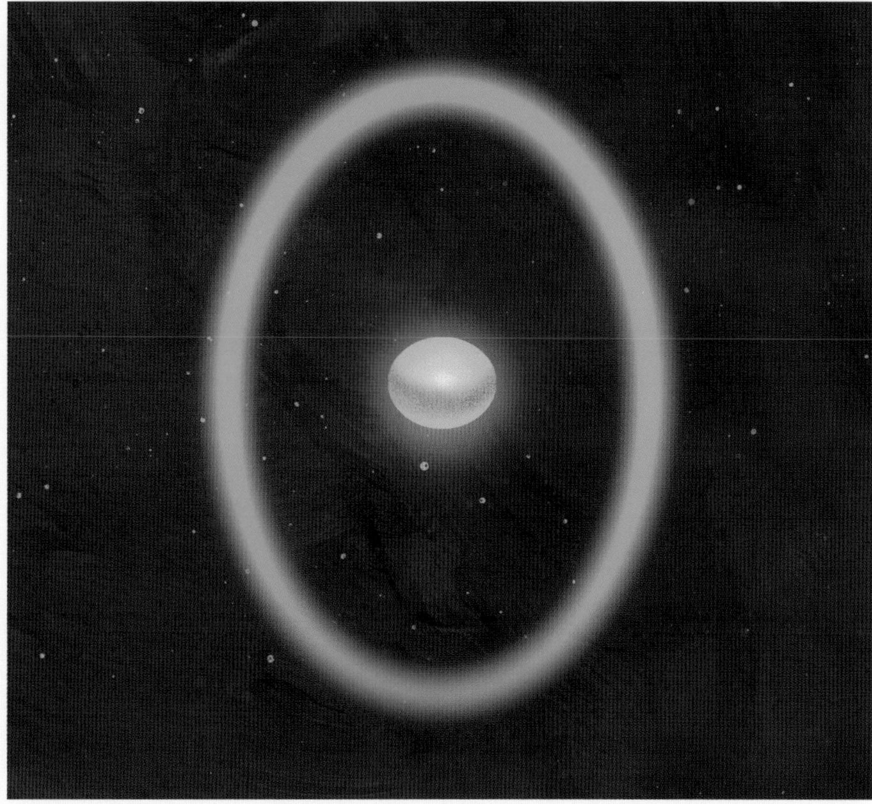

CRUMBLING WORLDS

Beyond a location in a balmy habitable zone, what are the prospects for life on a world circling one of these exotic stars? Conservative astrobiologists point out that A and F stars die more quickly, which raises potential problems for the genesis and flourishing of living biomes. But others respond that the few-billion-year lifetime of such stars is long enough to produce life. Earth life, they say, has been around for 4 billion years.

Below: *The habitability of an exoplanet is determined, in part, by its stellar company. The only confirmed life in the universe lives on a planet orbiting a G-type star, and that star is a lone one. But many stars congregate in multiples; double- and triple-member systems are more numerous than single-star systems. Systems with multiple stars in close proximity, such as this contact binary (two stars that touch each other), may be unlikely candidates for life-sustaining planets.*

Right: *The ancient planet PSR B1620-26 b, nicknamed "Methuselah" or "the genesis planet," formed less than 1 billion years after the birth of the universe. It circles its double-star system once each century at the distance between the Sun and Uranus. Although a safe distance from its white dwarf, the planet is bathed in deadly radiation from the nearby pulsar.*

Above: *Kepler and other observatories have found many Hot Jupiters in tight orbits around their stars. The majority orbit red dwarfs, but there are others too—and the ones that get too close begin to lose their atmospheres in prodigious amounts. Their atmospheres may extend for millions of miles behind them in comet-like tails of gas and dust.*

KELT-9 b is not the last of the "nightmare worlds," and most scientists have favorites. Many of the less inviting exoplanets orbit life-destroying stars, such as the exotic planet nicknamed "Methuselah." The planet is extremely ancient and orbits two closely linked stars (a white dwarf and a pulsar). Waves of radiation scour the surface of the planet, and aurorae may crown its poles. It lies in a globular cluster, a globe of ancient stars 12,400 light-years away that may be up to 12.7 billion years old. Because of this, the popular press has dubbed the planet Methuselah. The planet is 2.5 times as massive as Jupiter and takes a century to make one circuit around its sun. Because of its distance, its temperatures probably dive to −200°C (−330°F).

For scary places, the planet Gliese 3470 b gets the vote of NASA's Eric Lopez. A vast cloud of material enshrouds the sub-Neptune-size planet, and its light tells us something strange is going on there.

Above: *In addition to its exotic ices, Gliese 436 b breaks the mold in another way: it lacks methane. Computer models suggest that giant worlds ruled by hydrogen (as can be seen in our own solar system) and having temperatures in the neighborhood of 725°C (1,340°F) should have large amounts of methane. Gliese 436 b has temperatures close to that, but the Spitzer Space Telescope revealed that this planet's air does not have methane. The finding demonstrates the diversity of exoplanets and the surprises still lurking in their study.*

"Most of the reflection is from dust, not fog," says Lopez. The tail of this planet is less comet and more gravel yard. Gliese 3470 b appears to be a member of a new class of planet, Lopez goes on to suggest. "We've found several examples of things that are more like disintegrating rocky objects. They are so hot you begin vaporizing and blowing off rocks. So you get dirty rocky comets. There is no gas, no water. You get variable light curves because of all the boulders and dust and chunks of materials. The tidal stresses are huge."

Researchers have seen several disintegrating worlds. The first was Gliese 436 b, a Hot Neptune that drags an extended tail behind it and puts off more heat than it takes in from its red dwarf star. A unique feature of Gliese 436 b is that it is made of ice which may be "burning." Planets this size have super-pressure ices unlike anything experienced on Earth, thus Gliese 436 b is doing a

slow burn of gases that would otherwise be frozen. The planet may have formed farther out in the cold outer regions of its planetary system, then migrated inward. There, it smolders today, dragging its comet-like tail.

At least two of these disintegrating planets have been found circling Sun-like stars. A few similar objects have been spotted around white dwarf stars, but this small handful of planets represents a tiny

Below: *Typical light curve of a planetary transit (above) and the approximate light curve of a disintegrating planet (below). The bowl-shaped curve is warped to one side because material in the tail partially obscures the starlight as the planet drags its debris behind it.*

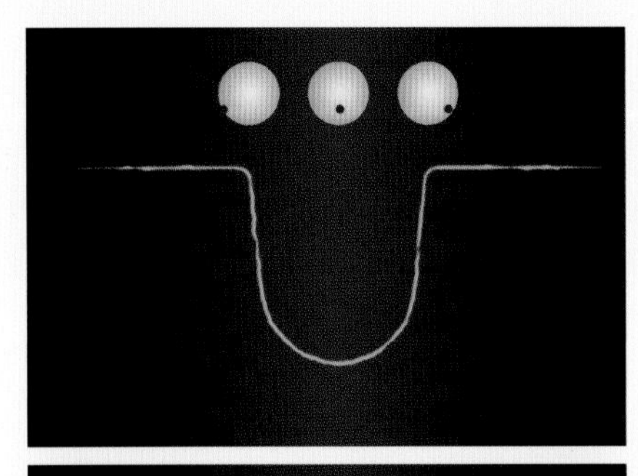

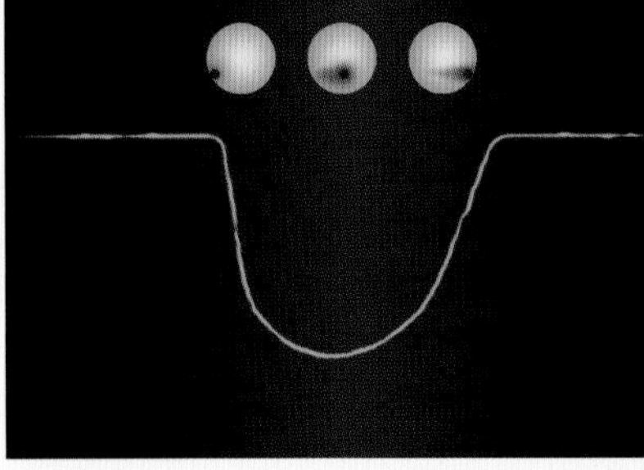

Above: *Heat from a dying core glows in the hollows of K2-22 b. What's left of the rocky Super Earth falls apart before a powerful torrent from its close-in star. The exoplanet is over two times the diameter of Earth, making it a candidate for either a rocky terran Super Earth or a gaseous sub-Neptune. But its diameter and mass imply a terrestrial nature. Disintegrating planets have been found at red and white dwarf stars, as well as main sequence stars more comparable to our own Sun. A terrifying world like this probably never got the chance to host life—unless a cataclysmic event sent it from a more distant, benign orbit into the deadly neighborhood nearer its star.*

minority of the nearly 5,000 exoplanets found so far. This shows astronomers that such planets must either be rare, or they must be short-lived.

COMETS OF STONE

K2-22 b imparts another tale of a tail. It is less than 2.5 times as far across as Earth and is probably more rocky than gaseous. The exoplanet drags a dusty trail behind it, but oddly, K2-22 b also has a tail of dust extending in front of it, entrained by the gravity of its small star. (Some comets display a double tail like this as well.) The planet intrigues Padi Boyd, head of NASA Goddard's Exoplanets and Stellar Astrophysics Laboratory:

"K2-22 b is falling apart. We may be able to get spectrum of the tail, which is the actual surface of the rocky planet." Being anywhere near K2-22 b is a treacherous prospect. The ground beneath your feet would continually shift and collapse and its skies are filled with the glowing dust that was once the landscape. But no air would tint the sky a comforting hue; only the blackness of space would loom overhead, tinged with powdered stone and rocky debris. Sheets of rock would peel off and blast into the sky, fragmenting into gravel and sand, joining the clouds of fractured rock above. On K2-22 b, stone becomes sky, and the only rainfall is the violent storm of sarsens and minerals.

Avi Mandell says that pinning down the true size of such worlds is difficult. "We don't actually know the real 'radius'

Above: *Super Earth LHS 3844 b has little or no atmosphere. It is a bleak, rocky planet with a surface similar in composition to the Earth's Moon or Mercury. Both the Moon and Mercury are covered in craters, but LHS 3844 b's size indicates that it may have retained enough internal heat to fuel volcanoes. Volcanic activity on a global scale may have resurfaced the entire planet many times over, leaving a relatively smooth, black landscape.*

THE GROUND . . . WOULD CONTINUALLY SHIFT AND COLLAPSE AND THE SKIES THERE ARE FILLED WITH THE GLOWING DUST THAT WAS ONCE THE LANDSCAPE.

Below: *The lava plains of Mercury (left) and the Earth's Moon may resemble the surface of LHS 3844 b. Here, the cratered face of Mercury has been wrinkled during the planet's cooling process. In the view of the Moon (right), layers of hardened lava peer from the lunar dust on the wall of Bessel crater.*

of K2-22 b. All we know is the total area blocked by the planet's material (that's where we get approximately 2.5 Earth radii, if you assume it's centered in a disk). But because of the tail, we think that the 'coverage area' is probably dominated by a dusty halo, and the planet's solid surface is much smaller. Similarly, the fact that so much material is escaping means that the surface gravity is low, and therefore the mass is low—possibly Mars-sized or smaller."

It has been difficult to get direct readings from the surface of any exoplanets, so collapsing worlds like K2-22 b and Gliese 3470 b promise to provide precious insights. But another exoplanet has given up some of its skin-deep secrets. LHS 3844 b is nearly Earth-size, with a mass of about 1.3 times that of our world. It orbits an M star 48.5 light-years away. Scientists turned the Spitzer Space Telescope on LHS 3844 b after the exoplanet was discovered in 2018 by the TESS spacecraft. They found a barren world devoid of atmosphere. The Spitzer sees in the infrared, or heat, part of the spectrum and it was able to essentially take the day and night temperatures of LHS 3844 b. The spacecraft found a huge difference in the two hemispheres, showing that no air was circulating to

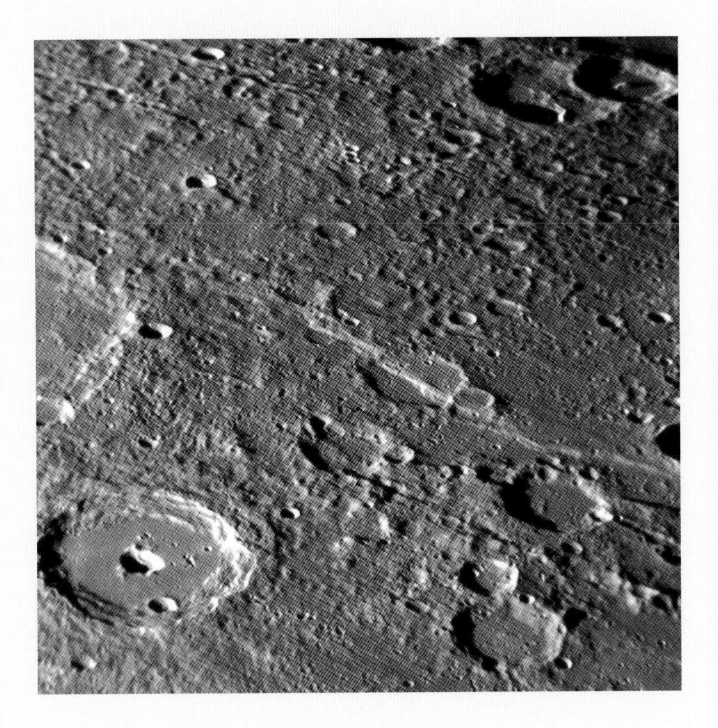

carry hot air to the night side and cold air to the dayside.

Researchers have compared the planet to the Moon or Mercury. In fact, on Mercury—closest planet to the Sun—daytime temperatures reach 425°C (800°F), but darkened areas are so chilled that ice remains in the shadows of craters at the poles. Temperatures in the shadows and on the nightside are frigid compared to the sunlit terrain, as no air evens them out. A similar situation seems to visit the even hotter LHS 3844 b, where daytime temperatures reach 770°C (1,420°F). The light coming from the surface is dark, similar to that reflected by the Moon's ancient lava plains called maria. Those dark lunar plains are made of basalt, as is much of the surface of Mercury. Basaltic regions like these were left by the activity of past volcanoes, so researchers at the Harvard and Smithsonian Center for Astrophysics in Cambridge, Massachusetts, suggest that LHS 3844 b may have a similar surface. It is likely an airless globe of broken lava and dark stone mountains, valleys, and craters.

SURVIVORS OF A STAR'S DEATH KNELL

While we have discovered a class of exoplanet that is disintegrating, and another class that is being stripped of its atmosphere, there is a third related group: planets that have survived the close-stellar-encounter process and lived on to tell the tale. A classic example is the exoplanetary duo of Kepler-70 b and c, which orbit the dying star Kepler-70. Kepler-70 is a former red giant that has blown much of its shell away. All that remains is a sun about one-half the mass and one-fifth the diameter of our Sun. Classified as a B-type subdwarf star, it burns six times hotter than the Sun. Its two known planets may have been close enough to have skimmed the outer fringes of their star as it swelled to a red giant, but they somehow survived destruction. Both may have been gas giants that lost their atmospheres to the furious stellar winds of their expanding sun, leaving behind the remnant cores of much larger planets. Kepler-70 b is three-quarters as far across as Earth and weighs in at about one-half the mass. Kepler-70 c has a high enough density that it may be composed mostly of metals. Both planets are among the hottest exoplanets known, with surface temperatures probably exceeding the surface temperature of the Sun. The sibling worlds complete their breakneck circuits around Kepler-70 in just 5.8 and 8.2 hours.

Planets like Kepler-70 b and c may be rare, but they would explain some odd formations in planetary nebulae. A planetary nebula is the leftover cloud of gas ejected from a dying star like a red giant. Gas giant planets embedded

Right: *A once vibrant planet lies cold and dead in the aftermath of its sun's collapse. A similar fate awaits most planets as their stars age and die. But will life on those worlds advance enough to escape to another habitable world?*

ANY SUCH PLANET AND MOONS, IF THEY HAD EVER HOSTED LIFE, WOULD BE LEFT AS STERILIZED, DEAD GLOBES.

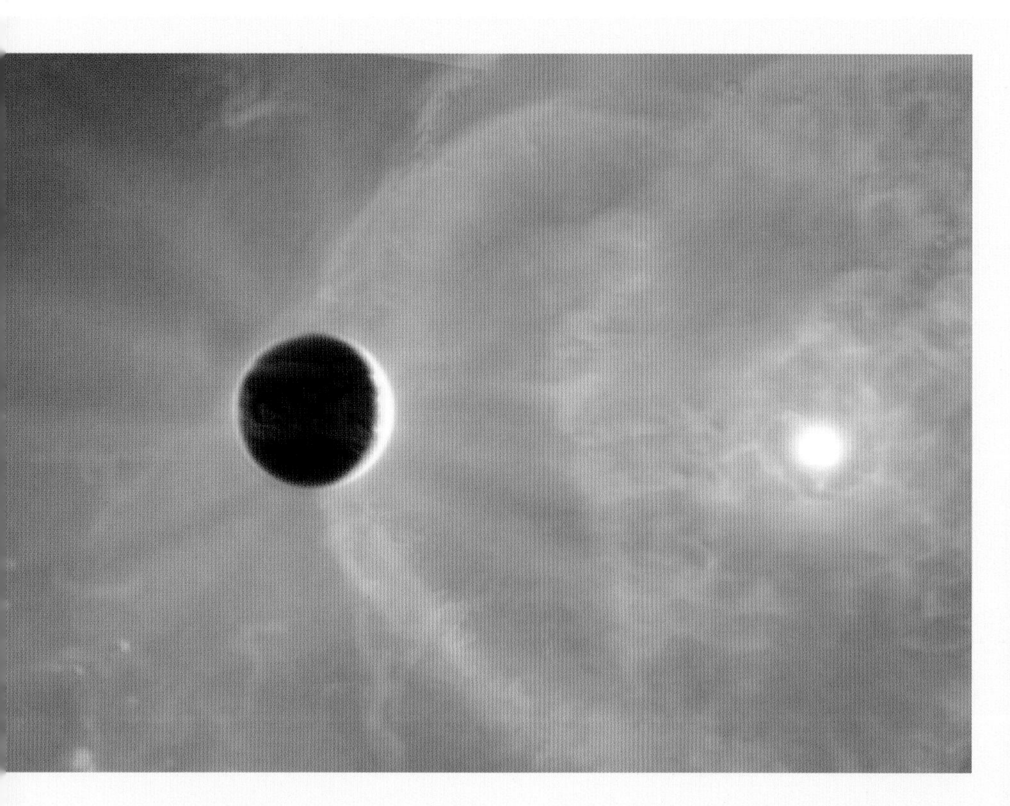

Left: *A planet originally several times Jupiter's mass has avoided being engulfed and destroyed during the death throes of its star. The outer layers of the star have been blown off to expose the intensely hot stellar core. Heat and radiation from the remnant core drive rogue waves of material from the extended atmosphere of the gas giant, while the distant, previously ejected shells of material from the star glow in the background as an emission nebula energized by UV radiation from the core.*

within newly formed planetary nebulae—something which has been written about but not yet directly observed—would lose an extensive amount of gas, resulting in a smaller gaseous planet. The star itself would collapse into a white dwarf or, if it began as a larger sun, a denser pulsar with beams of X-rays blazing from its poles. Any such planet and moons, if they had ever hosted life, would be left as sterilized, dead globes.

Giant planetary companions in tight orbits around their star may explain bizarre features in planetary nebulae such

Right: *The blast furnace world of Kepler-70 c seems in limbo between planet and star. Its surface—likely rocky in makeup—is on the point of incandescent vapor, with temperatures surpassing those on the surface of the Sun. Its planetary sibling, Kepler-70 b, is nearly lost in the glare of Kepler-70. The little exoplanet suffers even higher temperatures in a tighter orbit around its star. Kepler-70 b and c follow paths that bring them within 240,000km (150,000 miles) of each other, the closest encounter between exoplanets yet found.*

Above: *Not all planets make it to a stage where life can take hold. Some are ejected from their system by the gravitational billiards game of worlds circling stars. Some cross the orbits of other planets, colliding into a circumstellar ring of debris, an asteroid belt, or a ring around a giant world.*

as twisted gases, knotted clouds within disks, and springlike formations around polar jets, destined to drift eternally in the darkness between the stars. As these planets plow through the material surrounding the star, they can spin up the stellar cloud so rapidly that material is ejected, forming extended disks or tori around the star. If a planet spins up the material more gently, inner regions of the cloud will spin faster around the star than the outer reaches of the cloud, twisting the gases up into beams shooting from the poles. In a third case, the planet is actually pulled in toward the star, then rocketed out of the system on one of the polar jets. Kepler-70 b and c may be gearing up for such a ride.

DECEPTIVELY LOVELY: TAKING THE TEMPERATURE OF A PLANET

Some planets give the appearance of a benign environment, but these deceptive worlds will carry no life. One of the first exoplanets to reveal its true colors was the Hot Jupiter HD 189733 b. Preliminary work suggested that the planet had a blue tint, but results were inconclusive until the Hubble Space Telescope took a look. Hubble was able to observe the planet's face in visible light and it found a rich, blue world.

This second "pale blue dot" is inviting from afar, but its true nature is not so friendly. It is the closest Hot Jupiter yet found, so it has provided much information about Hot Jupiters as a class of planets. HD 189733 b was the first planet to be mapped thermally; researchers were able to create a "heat map" using the Spitzer Space Telescope. Circling its K-type orange dwarf sun in a tight orbit, daytime temperatures on the planet skyrocket to nearly 1,200°C (2,200°F). But the atmosphere flows across the planet, mixing hot daytime gases with the cooler evening air. Spitzer showed nighttime temperatures dropping by 400°C (720°F). This difference in temperature must set up raging, supersonic winds spiked with shards of glassy silicate crystals, flying microscopic blades. It is the silicate—usually found in the form of solid rock—that tints the atmosphere blue. Despite its inviting blue skies, HD 189733 b joins the ranks of extreme worlds where organic material simply cannot exist; there is no life on this superheated Hot Jupiter.

Taking the temperature of a planet 65 light-years from Earth isn't easy, and astronomers had to tease the planet's light out of the starlight's glare. Spitzer observed the star and its planet for 33 hours; the observing run began when the nightside of the planet was facing Earth. As the planet moved halfway around its orbit, a widening slice of the dayside appeared. The light of the star—when the planet's nightside was contributing no light—was subtracted from the light when the planet's face was reflecting light back to Spitzer. What was left was the planet's light, a hue made blue by floating vapors of molten silica, similar to melted glass.

Right: *The clouds of HD 189733 b fan out from the sunlit side in winds reaching 8,700km/h (5,400mph). If this searing Hot Jupiter ever had a ring, it was short-lived: rings do not last around planets that are tidally locked to their stars. Hot Jupiters may commonly lose their moons as well. Any natural satellite would fracture into a short-lived ring. Like K2-22 b and other close-in planets, HD 189733 b is so close to its sun that its atmosphere is streaming away at high rates.*

Below: *Jupiter (right) and HD 189733 b (artist concept at left). The blue Hot Jupiter is slightly larger and more massive, with a mass 1.16 times that of Jupiter. In contrast to Jupiter, temperatures on HD 189733 b are hot enough to melt silicates.*

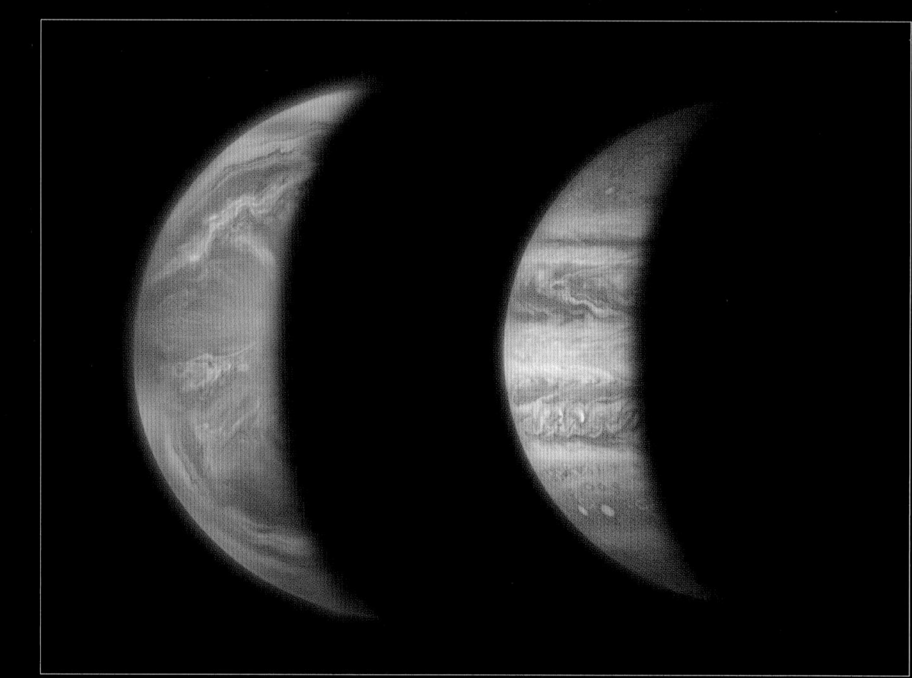

DEADLY SUPER EARTHS

In our search for life-friendly worlds, more planets with terrifying countenances await us throughout the cosmos, and some fall into the class of Super Earths. One of them, LTT 1445A b, is just half again larger than Earth, but it packs a punch: the planet holds 2.2 times the mass. Its weighty nature means that the planet is likely a rocky one.

At a distance of 5.7 million km (3.5 million miles), LTT 1445A b is so close to its star, a red dwarf, that it is probably nearly devoid of atmosphere. It makes each circuit in just five days, but it has spectacular skies blessed with three suns. Its star, LTT 1445A, orbits two other red dwarfs in a triple-star system. But the planet is no place for a picnic: LTT 1445A b loops around its star so closely that temperatures there are as hot as a bakery oven, roughly 150°C (300°F).

One planetary system was doomed from the outset by a behemoth planet in an odd orbit. Pi Mensae is a G star just slightly larger than our Sun, and when preliminary results suggested the presence of planets, astrobiologists were excited to see what was in orbit around

ANOTHER SUPER EARTH HAS NOT BEEN SO SECRETIVE ABOUT ITS WATER, AND ITS DISCOVERY IS AN EXCITING ONE IN OUR SEARCH FOR LIFE-BEARING WORLDS.

Left: *The Super Earth LTT 1445A b—discovered by the TESS orbiting telescope—simmers well inside of its star's habitable zone. Any visitor would be rewarded with spectacular views of occasional double sunsets of the two distant stars in the system, LTT 1445B and C (seen at upper right). Stars B and C orbit closely around each other, while A follows a path outside of them both.*

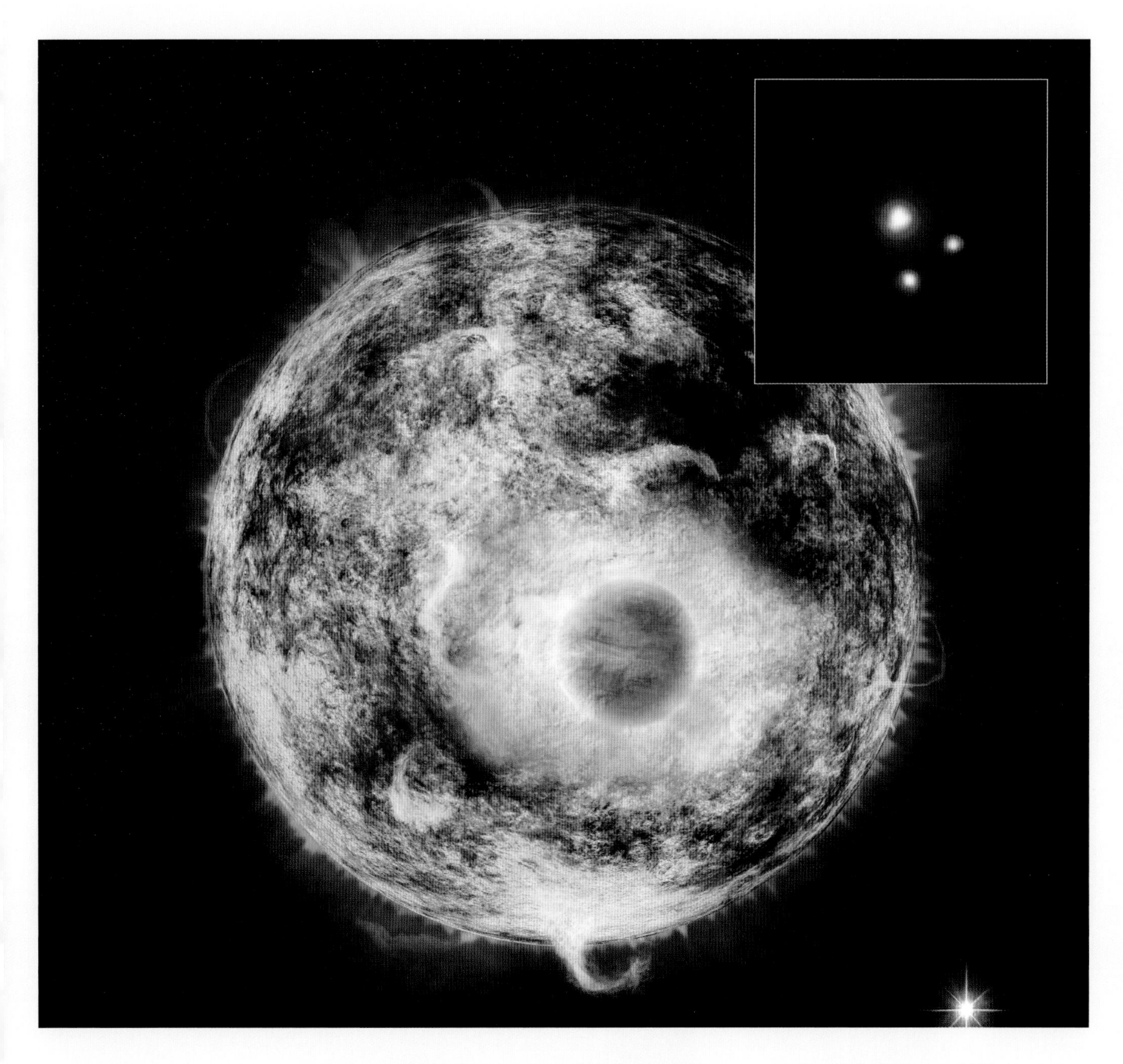

Above: *An artist's concept of another searing world in orbit around the Sun-like star HD 189733. This view shows the star transited by the exoplanet HD 189733 b. The exoplanet is a Hot Jupiter roughly the size of Jupiter with surface temperatures comparable to the open flame of a candle. A faint red dwarf companion star glows in the background. This star circles the main star in a long, distant orbit lasting 3,200 years. The inset contains the Chandra X-ray Observatory image of HD 189733. The source in the middle is the main star and the source in the lower right is the faint companion star.*

GAS GIANTS IN HABITABLE ZONES: THE MOON FACTOR

Planetary dynamicists calculate that for any moon of the planet to be in a stable orbit, it would require an orbital period lasting no longer than about four months. In the case of moons nearer to the giant planet, or moons in orbits close to sibling moons, gravitational influences could easily trigger tidal heating within the moon, sustaining geological activity. These volcanoes can replenish atmosphere and offer havens for mineral-munching microbes. Volcanism, combined with the presence of liquid water in the habitable zone, makes moons of Upsilon Andromedae d prime candidates for life.

Right: *The gas giant Upsilon Andromedae d lies in the habitable zone of its star. It is likely that the giant planet has some kind of ring system, since all the giant planets in our solar system do.*

this Sun-like star. What they found was a gigantic planet hell-bent on destruction. Pi Mensae b is one of the most massive exoplanets ever found. But the oversize world is a planet-killer because of its orbit. Pi Mensae b follows an eccentric path, taking it on a 2,151-day course around its star in an egg-shaped trek. This orbit brings the monster through the habitable zone, but it doesn't linger long enough for life to take hold. Instead, the planet's intense gravity has probably tossed any habitable zone planets from their orbits, sending them into a fiery death in their sun or condemning them to a frigid demise in the outer darkness.

Pi Mensae b has company in the eccentric orbit department. The Upsilon Andromedae system is a binary-star system whose main sun is an F star, while the other is a red dwarf. The planetary family orbiting the primary star has at least four worlds and was the first multiplanetary system discovered around a main sequence star. Its two outer planets, Upsilon Andromedae c and d, follow orbits among the most eccentric yet seen. Planet d lies within the massive star's extended habitable zone.

Our search for life-bearing Super Earths brings us to Gliese 1214 b, a misty world 2.5 times the size of Earth, but just one-third as dense. Its mass and size provide a range of possibilities for the planet's makeup. It is possible the Gliese 1214 b has a rocky core embedded in a deep hydrogen/helium atmosphere. But the Super Earth could also be

Solar systems don't need renegade supergiants like Pi Mensae b to toss planets from their habitable zones.

a steam planet, with a water vapor canopy covering a deep global ocean. Using a horde of ground-based and orbiting telescopes, researchers set out to uncover just what the planet is like. Their observations demonstrated that the exoplanet's atmosphere is an effective barrier to its sun's starlight. Astronomers expected to see the strong signature of water, but they did not. Instead, their combined efforts showed an absence of water vapor. They knew that in a planet this size, there must be some water present. What was masking it? The answer was clouds.

NASA exoplanet modeler Thomas Fauchez has a love-hate relationship with clouds. "The warmer the planet, the more clouds you tend to have. The more water you have, the more clouds you get, but those clouds hide the atmosphere." In the case of Gliese 1214 b, clouds were short-circuiting any attempt to find the water, but atmospheric scientists had other tricks up their sleeves. They were still able to measure materials in the atmosphere—materials that can tell us about how a planet handles incoming and outgoing heat, what its chemistry is like, and how the constituents in the air interact with each other. And while our observatories were able to tease out some ingredients in the skies of Gliese 1214 b, water is among the compounds that tell us the most—along with hydrogen and carbon dioxide—and it is the one compound we cannot see there. This has been frustrating to scientists. Hubble is good at finding water and has found its presence in nearly eight out of 10 large exoplanets. But future missions like the James Webb Space Telescope will be able to solve the mystery of just how much water floats in the atmosphere of planets like Gliese 1214 b.

Another Super Earth has not been so secretive about its water, and its discovery is an exciting one in our search for life-bearing worlds. Planet K2-18 b, an exoplanet located 110 light-years away, possesses moisture-filled skies. This discovery of water vapor—and perhaps liquid water droplets—in a Super Earth atmosphere was the first; no other exoplanet had been confirmed to have conditions favorable for liquid water. The planet lies in its star's habitable zone, but like most Super Earths, its exact nature is unknown. K2-18 b may have a rocky surface, but it is more likely a giant world of liquid and gas, similar in consistency to Neptune. Models and new data indicate that K2-18 b may have the right combination of temperature and pressure to accommodate rainfall in its skies. But the planet is over twice the diameter of Earth and eight times its mass. Although it is an alien world unlike Earth, the discovery of a planet with liquid water, a key necessity for life as we know it, adds another candidate to our list of planets that may support life.

SHIFTING ZONES AND LIFE'S OPPORTUNITIES

Solar systems don't need renegade supergiants like Pi Mensae b to toss

Above: *Super Earth Gliese 1214 b orbits the nearby red dwarf Gliese 1214. Most models predict that the exoplanet is an ocean world, with a deep hydrogen/helium/water atmosphere resembling that of Neptune. Water would not congregate into conventional oceans, but rather transition from vapor into a super-pressure form not seen on Earth. The exoplanet weighs six times the mass of Earth.*

planets from their habitable zones. Sometimes stars strand planets on their own. As a star matures, its habitable zone shifts. Work by astrophysicist Michael Hart reveals that the Sun's habitable zone began closer in and has migrated outward over time as the Sun has aged and grown brighter. Four billion years ago, during the Earth's formative years, the Sun put out only 70 percent as much heat and light as it does today. Our star continues this trend, forcing our habitable zone outward. Today, the Earth is in a central enough location that it will remain in the habitable zone throughout this shift. But had the Earth formed just 1 percent farther away, our world would have suffered a runaway glaciation. If the Earth had been born a few percentage points closer to the Sun, we would have suffered a runaway greenhouse effect similar to that seen on Venus. The habitable zone is a planetary tightrope.

In the same way, if the Earth's orbit were a bit more out-of-round than it is, its route would take it out of the habitable zone as time steadily progressed. But the flip side is that a frozen planet on the outer edge of the zone may become habitable as its star ages and warms. Habitable worlds closer in may become too hot to support life as their stars brighten.

THE DOWNSIDE OF AN EGG-SHAPED VOYAGE

One factor in the habitability of a planet is the shape of its orbit. A circular orbit brings regular cycles of day/night cycles and—especially—seasons. But an elongated orbit can cause problems for life. Take the case of HD 80606 b, a gas giant orbiting a G-type star. Its orbit is one of the most eccentric known, ranging from a distance equal to that between Earth and Sun at its farthest, down to a horrifying three-hundredths of that. NASA Goddard's Avi Mandell outlined the planet's intriguing case. "For the few days that the planet is nearest the star, it's being hammered by the star, so you get this very bright dayside and very dark nightside, but as the planet circles away now the planet is circulating that energy." As the planet pulls away from its star, its atmosphere mixes, cooling the planet. "You start with a dichotomous day to night, but by the time you're down in the orbit you get this mix, with dramatic vortices and swirls. It mixes because you're not being heated anymore." During the closest point of its 111-day orbit, HD 80606 b's dayside temperatures soar to 1,225°C (2,240°F). Mandell adds that "It's not tidally locked, but we have no idea what the spin rate is. In fact, the whole extended orbit is unknown because we can't see it."

Below: *At the farthest point of its orbit (aphelion), the gas giant HD 80606 b sees a sun in the sky similar to the Sun's appearance from Earth. But as it makes its way to the nearest point in its path (perihelion), temperatures soar and the star swells to a glowing terror.*

Stable regions within habitable zones of cooler M and K stars, whose lifetimes are drawn out, last longer than those for short-lived suns because the habitable zones migrate over a longer period of time.

Climate controls appear to be built into the design of Earth. The most influential of these dampers is a chemical recycling process called the CO2/rock cycle. Carbon dioxide (CO2) is a powerful greenhouse gas and without it the Earth's surface would average 40°C (100°F) colder than it is, because the CO2 acts as an insulator, holding the heat in. In a complex feedback loop, the CO2/rock cycle steadies global temperatures. As Earth warms, increased weathering removes CO2 from the air. That drop in greenhouse gas levels lowers global temperatures. As the temperatures sink, the weathering processes subside, allowing the CO2 levels to rise again. A second stabilizing activity depends on plate tectonics. The rocky surface of the Earth acts as a chemical sponge, soaking up various gases and chemically lodging them within the rocks. The Earth's plate tectonics bring the gas-soaked rock down into the mantle where it melts, freeing

Left: *Exoplanet 47 Ursae majoris b is a super-Jovian planet 2.5 times the mass of Jupiter. The planet is outside of its star's habitable zone and has probably disrupted formation of large terrestrial worlds within that zone. Here, an artist shows a view of the planet from a hypothetical moon.*

the chemicals, gases, and minerals. These materials are then free to recycle back into the environment through mountain uplift or volcanism.

Taking into consideration the CO2/rock cycle and other moderating factors, atmospheric experts estimate that the Sun's habitable zone spans a distance of 0.95–1.15 AU from the Sun. It is a thin, precarious ring of migrating cosmic territory where life can take hold. Earthlike planets with surface conditions that remain habitable over long periods may be rare indeed.

With the discovery of large planets in habitable zones, the chance for life among their moons has become a game changer. Super Earths or gas giants have enough gravity to hold retinues of planet-size moons akin to Titan or Ganymede, and these moons will have plenty of raw materials for biological processes. The concept was even embraced by Hollywood. James Cameron's movie *Avatar* takes place on a very Earthlike moon of a large blue gas world.

Above: *Although giant gas worlds may not hold much promise for life, their moons may. Here, the artist imagines a cratered moon with an atmosphere and water-filled hollows.*

Right: *One simple yardstick for seeking out life across the galaxy is the location of a planet. It does not take much distance outside of the habitable zone to make a frigid world like that of OGLE-2005-BLG-390L b. The planet, nicknamed Hoth in a nod to the* Star Wars *planet, is 1.7 times the diameter of Earth. With surface temperatures of –220°C (–370°F), life there is unlikely.*

. . . GAS GIANTS HAVE ENOUGH GRAVITY TO HOLD RETINUES OF PLANET-SIZED MOONS AKIN TO TITAN OR GANYMEDE.

AMAZING DISAPPEARING PLANETS

Many uncertainties plague exoplanet hunters. Transiting planets can be disguised in the uneven glow of their stars, or they may hide within the periodic patterns of other transiting worlds. The planets within the habitable zones of Sun-like stars cross their sun's face only once each year or so, making it difficult to track them for long periods. Some planets cross the face of a star only once, or tug on their sun—shifting its starlight—but then seem to disappear.

Searches for exoplanets require delicate and complex data, and that data must be confirmed over long periods of time. The history of exoplanet hunting is littered with cautionary tales of planets that have come and gone. One such planet is HD 188753 Ab. Initial indications were that the hot gas giant—more massive than Jupiter—orbited the star HD 188753 A in just over three days, circling only 8 million km (5 million miles) away. Its sun is part of a triple-star system. It seemed unlikely that such a huge planet could survive at a distance less than one-twentieth of that from Earth to the Sun. And astronomers wondered: Why was it even there in the first place? Models suggested that planets could not form in a system like this one. Large planets were thought to coalesce in the outer reaches of a young star's formative cloud, but that cloud would be disrupted by the other nearby stars. HD 188753 A's planet-forming disk of dust and gas would have ended around 1 AU from its star. A huge gaseous planet should not have been able to form under those conditions.

Doubters seemed to be bolstered when attempts to confirm the planet were unsuccessful. In 2007, a team at the Geneva Observatory stated that they had the precision and sampling rate sufficient to detect the would-be planet, but that they did not detect it. Either HD 188753 Ab is not where predictions say it should be, or it was never there in the first place. Perhaps, as one astronomer quipped, it has simply vanished.

It would not be the only planet whose existence is in question. Observers at the Lick-Carnegie Exoplanet Survey announced the detection of a planet orbiting the red dwarf Gliese 581. The planet, Gliese 581 g, was part of a system of at least three planets, but some teams claim a fourth and fifth. Astronomers used two sources to identify Gliese 581 g, but other teams were not able to find it in their data. In 2014, a group of researchers asserted that the planet doesn't exist; but in 2015, a later team took another look and disagreed. For now, Gliese 581 g remains in the unconfirmed category. If the planet does exist, indications are that it is a Super Earth. It is joined by two other Super Earths, both confirmed: Gliese 581 c (which orbits at the inside edge of the habitable zone) and Gliese 581 d (which is embedded within the zone). Any moons that Gliese 581 d may have are good candidates for life.

Researchers estimate that Gliese 581 g probably has a radius no larger than 1.5 times that of Earth. Models applied to the proposed planet investigate factors

THEIR DISCOVERY DATA WAS NOT RELIABLE ENOUGH TO MAKE THE CASE FOR THE PLANET . . . GAMMA CEPHEI AB BECAME A CANCELLED WORLD.

Above: *Two views of Gliese 581 g: estimates for conditions on the planet range from a moderated Venus-like hothouse (left) to an icy Super Earth with a sun-warmed, chevron-shaped ocean and tropical conditions (right).*

like its atmosphere, size, and surface composition. Depending on the content of its atmosphere and surface, the planet may be a barren, Venus-like world. But if conditions are right and it is farther out in the habitable zone, the Super Earth may be quite a different place. If the air pressure is similar to the Earth's, models indicate that the world may be encased in a thick rock/ice crust. If the atmosphere is thin, global temperatures may range from –64° to –45°C (–83° to –49°F), but if the air contains enough greenhouse gases like methane and carbon dioxide, temperatures may rise considerably. Because of the fact that it is tidally locked, Gliese 581 g's atmospheric circulation may produce a permanent sea of liquid water. This body of water would face its sun just at the substellar point (with the sun directly overhead). Temperatures in this region could be as warm as those in the Earth's tropics. Studies of airflow show that wind patterns would lead to a great, chevron-shaped warm region pointing roughly along the equator.

Gliese 581 g provides yet another example of how it is theoretically possible for a tidally locked planet to support life. If a planet's distance and temperatures are similar to those of Gliese 581 g, or if it has an atmosphere that creates stable environments along the terminator (like the "eyeball" planets), conditions may be favorable for life even on such an alien world. If Gliese 581 g is confirmed, these factors combine to make it one of the most Earthlike worlds, with an ESI as high as 0.90. Not bad for a planet that may not exist.

Our last case study of an on-again, off-again planet is a gigantic world called Gamma Cephei Ab, nicknamed Tadmor. More massive than Jupiter, Tadmor's existence was announced in 1988. Canadian astronomers thought they had found the first exoplanet. But in 1992, the team announced that their data were not reliable enough to confirm the planet. The team cautioned that periodic shifts in the star's magnetic fields might have mimicked the gravitational pull of an unseen planet. They retracted their findings. Gamma Cephei Ab became a canceled world.

The cancellation may have been premature. In 2002, a new team of astronomers—including some of the original members—verified the existence of Tadmor. The team used multiple radial velocity readings. Gamma Cephei Ab now joins the ranks of confirmed exoplanets. Perhaps one day, Gliese 581 g will join it.

KINDER, GENTLER SUPER EARTHS?

Planets with glass-laden hurricane winds. Worlds as hot as the surface of the Sun. Globes of burning ice. A survey of Super Earths can be a disappointment for those seeking life in the galaxy. But all is not lost. Some Super Earths may, in fact, be excellent candidates for Earth 2.0. These inhabit the top of our A-list in the last chapter.

One such hopeful contender is Kepler-452 b, a Super Earth orbiting a Sun-like star in a very Earthlike orbit. Circling its sun every 385 days, the planet is nested deep within the habitable zone of Kepler-452. At over 1,000 light-years away, it is difficult to study, but researchers have gleaned clues to its nature. Kepler-452 b's mass is less than five times that of Earth, well within the constraints of a gaseous, water-vapor-ruled sub-Neptune. But researchers studying the planet's size and mass lean toward a terrestrial planet with a solid surface and likely dense atmosphere. With a diameter half again as large as the Earth, the planet likely has active geology, with

Below: *Comparison of Earth, at left, to Kepler-452 b, a Super Earth about 60 percent larger than our world. The exoplanet's size causes analysts to lean toward a terrestrial—rather than water world—interpretation of the planet.*

Circling its sun every 385 days, the planet is nested deep within the habitable zone of Kepler 452.

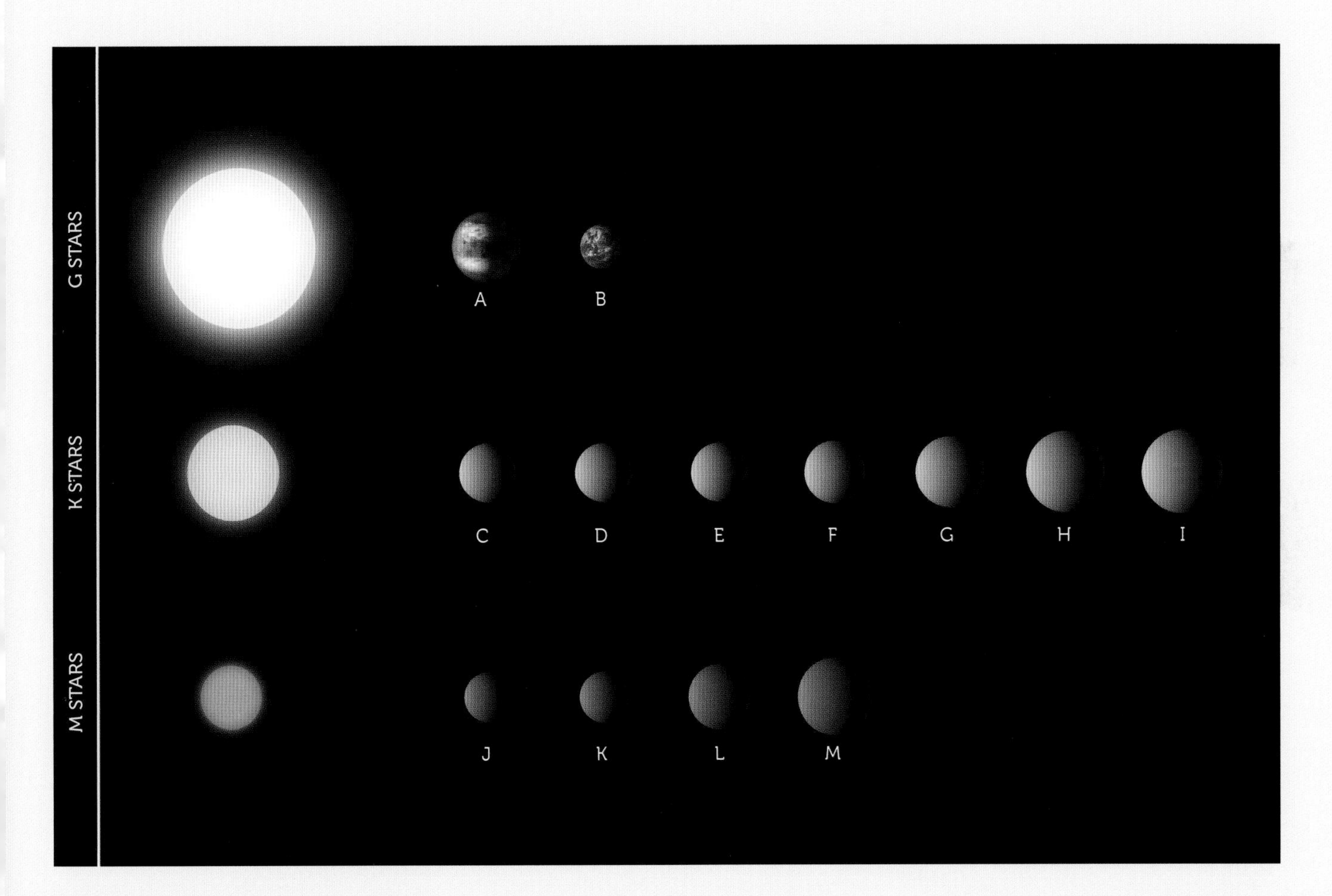

Above: *A dozen Earths: 12 planets less than twice the size of Earth, all within their stars' habitable zones. In this NASA diagram, planets are enlarged 25 times compared to their star. A=Kepler-425 b; B=Earth; C=442 b; D=155 c; E=235 e; F=62 f; G=62 e; H=283 c; I=440 b; J=438 b; K=186 f; L=296 e; M=296 f.*

frequent volcanism. Water may remain in vapor form if the atmosphere is thick, but models indicate that thinner atmospheres would support an active water cycle, with condensing clouds, rainfall, rivers, and seas. This surface water could buffer the environment from a runaway greenhouse effect in the atmosphere, as would the active geology, making Kepler-452 b one of the most Earthlike planets known. Kepler-186 f is another potentially inhabited world. Discovered by Elisa Quintana, a NASA Ames Kepler team member, the planet was the first Earth-size world to be discovered within the habitable zone of its star. It orbits Kepler-186 every 130 days. Daytime temperatures are estimated to average

around a chilly −85°C (−121°F), but some life on Earth thrives in similar conditions. Kepler-186 f's Earth Similarity Index is 0.58, about the same as Mars.

The planet's siblings all check in at less than half the size of Earth. Their orbits take them so close to Kepler-186 that they are far too hot for organic material to survive.

Counted among this privileged Earth 2.0 club is an exoplanetary duo we have visited before: Kepler-62 e and f. These water world terrans may well be among the planets most like home. Models vary in terms of which of the two planets sustains the most Earthlike conditions. The planets of Kepler-62 come with all the caveats that planets at red dwarfs must, like tidal locking and sporadic radiation spikes (see Chapter 3). But another, Kepler-442 b, has no such

Above: *Planetary systems of Kepler-186 and the Sun, compared. The scale of the diagram lets us see only the innermost terrestrial planets of the Sun's family (A, B, and C). Kepler-186's system (b,c, d, e, and f) fits neatly inside the orbit of Mercury, but because of the feeble heat of the red dwarf star, the outermost Kepler-186 f is in the habitable zone.*

Right: *An artist's concept of Kepler-186 f depicts the larger-than-life Earth as a rocky terran with water on its surface. This Earthlike version of Kepler-186 f has gained favor in the scientific community as more data come in. According to a NASA press release, Kepler-186 f is "the first validated Earth-size planet to orbit a distant star in the habitable zone."*

WATER MAY REMAIN IN VAPOR FORM IF THE ATMOSPHERE IS THICK, BUT MODELS INDICATE THAT THINNER ATMOSPHERES WOULD SUPPORT AN ACTIVE WATER CYCLE, WITH CONDENSING CLOUDS, RAINFALL, RIVERS, AND SEAS.

Residing within its star's habitable zone, Kepler-442 b is considered by many experts to be the most promising planet for habitability and life.

Left: *Kepler-62 is about two-thirds the mass and diameter of the Sun. Two planets orbit in its habitable zone, Kepler-62 e and f. Kepler-62 e orbits on the inside edge of the habitable zone. The planet may be a rocky Super Earth (pictured), but it is 1.6 times as far across as Earth. This puts it at the lower border of a Neptune-like gaseous world, a poor candidate in our search for life. If, on the other hand, it is small enough to be covered by global oceans, they will be liquid. In this case, a dense, warm atmosphere is a reasonable model.*

Above: *On the outer edge of Kepler-62's habitable zone lies Kepler-62 f, smaller than its inner sibling and probably just on the outer border where water freezes. It is likely a rock world covered by a global ocean; that ocean may well be frozen, at least on the surface.*

reservations. It orbits a K-type star just a bit smaller and cooler than our own, and it is far enough away to have escaped tidal locking. Its orbit is closer than the Earth–Sun distance and it is likely that its day is longer than Earth's, perhaps lasting weeks or even months, a significant portion of its 112-day orbit. Residing within its star's habitable zone, Kepler-442 b is considered by many experts to be the most promising planet for habitability and life.

A ROCKY SUPER EARTH?

Kepler-442 b is a planet on the brink. It is large enough that some researchers suspect it is gaseous, but close enough to Earth-size that others suggest it is an oversize rocky terran. Its size indicates that geologic activity is likely, which would increase the possibility of liquid water on the surface. The planet orbits within the habitable zone of the K-type star Kepler-442. It receives only three-quarters of the light and heat that the Earth gleans from its Sun, leaving Kepler-442 b a chilly terran with possible liquid water in various warmer spots across the globe. But because of its higher gravity, a dense atmosphere is also a possibility, lending greenhouse warming to its environment.

Above: *The Super Earth Kepler-442 b spans nearly 1.5 times as far across as the Earth, with twice the mass. These factors point to a rocky nature for the exoplanet.*

Top: *Kepler-442 b may be a colder version of the Earth, with most of the water lying frozen on the rocky surface. Models range from an Earthlike atmosphere to one more like the dense air of Venus.*

Above: *A starry daytime sky on the planet Kepler-16 AB b includes two suns. Like* Star Wars' *desert planet Tatooine, Kepler-16 AB b orbits two stars, one of them a red dwarf. But unlike Luke Skywalker's home, Kepler-16 AB b is a gaseous giant. The huge world probably has many moons and it is seen here from one of them. Could the moons of this world—extreme as they are—have life?*

Our survey of planets that would terrify everything from microbe to mammoth provides a key insight: Earthlike worlds are hard to come by (and ones with active, living biomes are probably even more uncommon). There may be a few, like Kepler-442 b and Kepler-452 b, but these may be the exception to the rule, rarities among billions of planets and moons. This reflects on our own Earth, a blue singularity, a globe to be cherished and cared for. But life may find a way, even in those extreme environments barely resembling Earth conditions, to take hold and flourish. What of those rare exceptions, those worlds that may stretch our concept of habitable environments or living systems? What riotous forms of life may thrive across the starry sky? We tackle that frontier in the next chapter.

5

IT'S AN ALIEN LIFE

Gene Roddenberry populated his *Star Trek* universe with a wide variety of aliens. Budget constraints dictated that most were variations of humans, with skin tinted odd colors or antennae sticking out from their heads. Even the silicon-based Horta appeared to be a stagehand lurking under a decorated carpet. George Lucas treated us to a similar menagerie of off-world inhabitants in *Star Wars*, especially in his Mos Eisley Cantina. Aliens—and our concept of them—became more sophisticated as budgets soared and science grappled with the great question posed by Enrico Fermi: "Where is everybody?" Some were terrifying, like the creatures in the *Alien* movie series or H. G. Wells's conspiring Martians in *War of the Worlds*. Perhaps our propensity for seeing extrasolar life as terrifying is our natural fear of the unknown. But others were far more benign and advanced, as witnessed by Arthur C. Clarke's *2001: A Space Odyssey*, Steven Spielberg's cuddly *E.T.*, Edmund H. North's guardians of the worlds in *The Day the Earth Stood Still*, or the time-jumping beings of Eric Heisserer's *Arrival*. But at the heart of a good story is a good conflict, and aliens provide natural fodder for such a plot device.

Left: *It may be that most advanced civilizations die quickly, exterminating themselves through their own technological advances. They will be outlasted by "lower" life-forms that have the common sense not to destroy themselves. Far from Clarke's beneficent alien races or Wells's jealous Martians, these simple survivors may represent the majority of life in the galaxy.*

LIFE AND THE DARK SIDE OF BIOLOGY ON EXOPLANETS

NASA Ames astrobiologist Chris McKay is definitely a fan of science fiction. When asked about the search for life in some corners of our own solar system, he paraphrases *Star Trek*'s famed physician, Dr. McCoy: "It's dead, Jim." But McKay doesn't think this is the case for all planets out there, or he wouldn't be in the business he's in. In fact, he is optimistic.

. . . WE SEEM TO BE SURROUNDED BY ENVIRONMENTS SIMILAR TO THE EARLY EARTH, WITH THE KEY INGREDIENTS OF LIQUID WATER, NITROGEN, AND CARBON.

McKay cites several reasons for his positive outlook. First, evidence for life on Earth has been found in the oldest sedimentary rocks on Earth, dating back to about 3.8 billion years ago. Life got started early and under harsh conditions.

Second, we seem to be surrounded by environments similar to the early Earth, with the key ingredients of liquid water, nitrogen, and carbon. Specifically, in our own solar system, we have the warmer, wetter early conditions on Mars and we have the current global seas of Enceladus. Lurking beneath the ice crust of Jupiter's moon Europa may also be an environment similar to Earth's deep oceans. "There are probably many hundreds of millions of exoplanets and exomoons that fall into the same class," McKay says.

The third reason for optimism concerns the elements of life, McKay says. "The four elements that make up 99.5 percent of the elements in life are hydrogen, oxygen, carbon, and nitrogen, and these are the first, third, fourth, and sixth most abundant elements in the universe, respectively. Note that numbers two and five are helium and neon and it's not surprising that life does not use them." McKay's fourth reason for optimism is that organic material is easily formed from carbon in a variety of environments. "We see this in meteorites, in the interstellar medium, and in the Miller-Urey experiments." Those experiments took a primordial brew of methane, carbon, and other elements, introduced radiation or electricity similar to lightning, and decanted a product of complex chains of organic material like amino acids.

Lastly, McKay points out that when life starts in one place, there appear to be ways it can be carried to other worlds through meteor impacts that blast life-bearing material into interplanetary space. These meteors eventually make landfall on other planets, as we saw with the Allan Hills meteorite. It may be that such journeys are possible from one star to another, with life in suspended stasis for the eons it takes to travel between solar systems. In many regions, stars are more densely packed than they are in our neighborhood, making such a delivery even more likely.

McKay concludes, "To quote Calvin to Hobbes. 'Let's go exploring'."

Below: *Some of the earliest traces of life on Earth are stromatolites, colonies of cyanobacteria layered with fine sand or silt. These early oxygen factories date back to nearly 3.5 billion years ago.*

THE EARLIEST ALIEN LIFE . . . ON EARTH

Some of the most ancient microfossils are complex one-celled microbes. Examples from the Columbia River valley have been documented with external sheaths and segmented membranes. A team in Greenland uncovered fossil evidence of structures left behind by ancient microbes in 3.7-billion-year-old rocks. Those rocks were from a shallow sea, while other primordial microfossils resided in superheated water of deep-sea hydrothermal vents. Life must have begun before these, and the two very different environments in which the fossils lived shows that life diversified early in the Earth's history.

How did life start on our planet? The question plagues biologists, but they understand, at least, the elements that it takes to make life. The early Earth's environs had no free oxygen in the air, presenting a truly alien environment. The Moon was closer, looming large in the sky with its seas of molten lava from great impacts of asteroids. Earth's atmosphere contained a mix of water, nitrogen, and carbon, all readily available to living systems. Lightning was in the air and the active Sun added its own influences, pouring radiation into all those carbon atoms. The energy from both sources triggered chain reactions in the chemistry of atmosphere and ocean, fusing carbon atoms with other materials to produce complex chains of organics, the building blocks of life.

Above: *Life on the primordial Earth got started early. The oldest microfossils—found by a British team in Canada—may date back as far as 3.8 billion years and were probably living on the ocean floor at a volcanic vent. Other organics rained from the sky, brewed by lightning or solar radiation.*

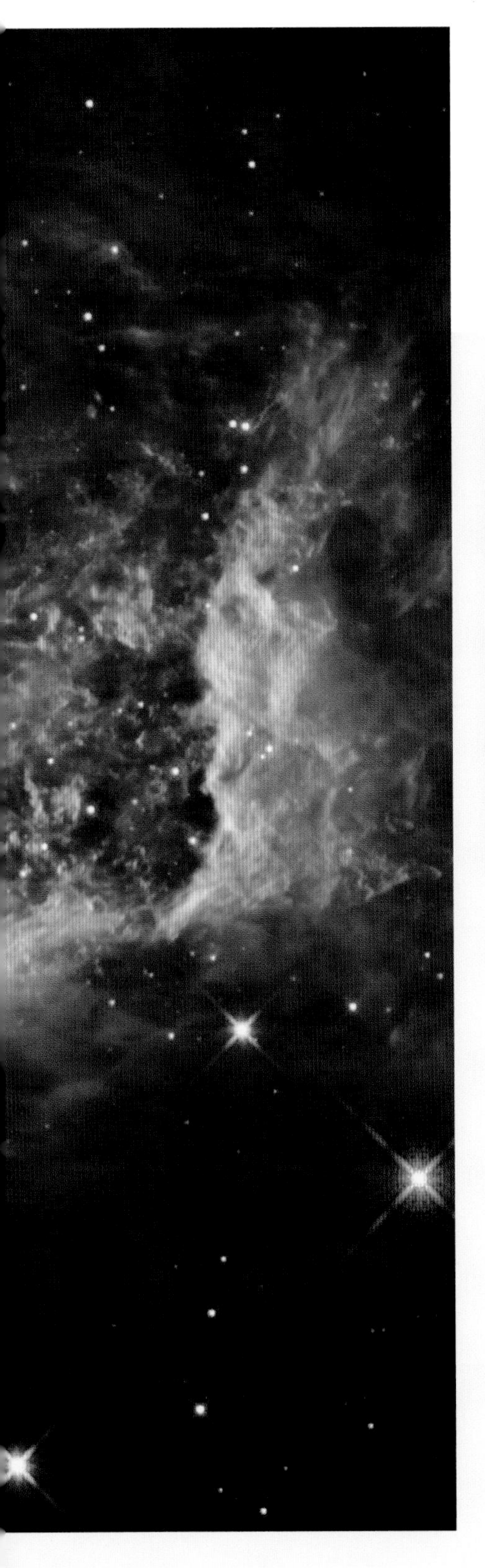

Above: *H. G. Wells penned* The War of the Worlds *to dramatize what might happen if evil-minded Martians invaded the Earth. His Martians, hell-bent on conquest of the Earth, made landfall in Woking, near London. Remakes of his story in film have placed the settings in other areas. Steven Spielberg's 2005 film took place in Brooklyn, New York, while the 1953 George Pal classic unfolds in southern California.*

Left: *The four primary elements used by life-forms within their metabolism are hydrogen, oxygen, carbon, and nitrogen. All four are found within the glowing gases of nebulae, where new solar systems are formed. Here, we see the star-forming nursery of Sharpless2-106, some 2,000 light-years away from Earth. The central area glows with the blaze of a young, active star within.*

We do not know whether intelligent life exists across the universe (even if we count ourselves), but if it does, will we find friend or foe? Will we find, as H. G. Wells had it, "intelligences greater than man's and yet as mortal as his own"? Or when we stumble upon life, will it be so fundamentally different from ours that we will not even recognize it? In his 1961 novel *Solaris*, author Stanisław Lem pens a global alien entity whose thoughts are so different from human intellect that any meaningful dialogue proves unattainable.

Another disturbing possibility is that advanced life always evolves to face a crisis point, and that many do not survive it. For humanity, that crisis may come in the form of nuclear or biological warfare, fatal powers that we unleash upon ourselves. Whatever the crisis, it may be that advanced civilizations across the galaxy are rare because they do not survive into an old age.

Most astrobiologists suggest that debating intelligent life may be getting ahead of ourselves. Simple and basic life-forms, some argue, are far more likely to be discovered than ones building shopping malls. Microbial creatures may not necessarily lead to complex ones as they may not survive the long-term changes in habitable zones, shifts in steady orbits, disturbances of stable environments, and extinction events like large impacts, all of which may cut short the progression of life from simple to complex. Microbial creatures or other simple life may be the common denominator throughout the exoplanets. Life may cling to the shorelines on planets within habitable zones, but it also may thrive in submarine environments on ice planets or moons.

Right: *Many worlds orbit at the fringes of their habitable zones, nearly too hot or barely warm enough to sustain liquid water. The unconfirmed exoplanet orbiting the star Beta Hydri has at least four times the mass of Jupiter and circles at a distant 8 AU, but its star is larger, brighter, and hotter than our Sun, so the planet's path lies at the outside edge of the habitable zone. Here, an Earth-size moon, easily possible with such a large planet, has a view of the Jovian planet as well as the bright Beta Hydri. Could life thrive under the ice or in the waterfalls of such a world?*

WHATEVER THE CRISIS, IT MAY BE THAT ADVANCED CIVILIZATIONS ACROSS THE GALAXY ARE RARE BECAUSE THEY DO NOT SURVIVE INTO AN OLD AGE.

HOW HARD IS IT TO MAKE LIFE?

The most common exoplanets, the Super Earths or sub-Neptunes, may not be so promising for life after all. Upon investigation, many turn out to be super-pressure steam worlds. Recent models indicate that planets the size of Earth or smaller tend to gather a dense cocoon of gases like hydrogen and helium early in their development, but they will lose their thick atmospheres later on, leading to more Earthlike conditions. But Super Earths appear to be a different matter: their massive gravity retains most of the hydrogen, creating a more Neptune-like than rocky world. Many Super Earths with low densities are likely sub-Neptunes.

Although the outlook for life on gaseous Super Earths may be dim, it seems likely that at least a fraction of the Earth-size exoplanets have seen the rise of complex biological forms. The majority of those planets orbit M stars. But astrophysicist Vladimir Airapetian has come to a haunting conclusion: the majority of exoplanets in the habitable zone are dead, nearly airless worlds. He looks to our own system as a guide. During the first 500–600 million years of the solar system's development, the

Below: *Close to a red dwarf sun, planets will end up with different fates. Earth-size rocky worlds (top row) will initially develop vast oceans, but these will be mostly blown into space by the action of the young M star. Later, as the red dwarf settles down into a more stable nature, the planet may retain smaller amounts of water and air, or these may be resupplied by comets and geological activity. Larger, Neptune-size planets (bottom row) will lose great quantities of water and atmosphere early on, but they will be large enough to retain deep, high-pressure global oceans as they become Super Earths.*

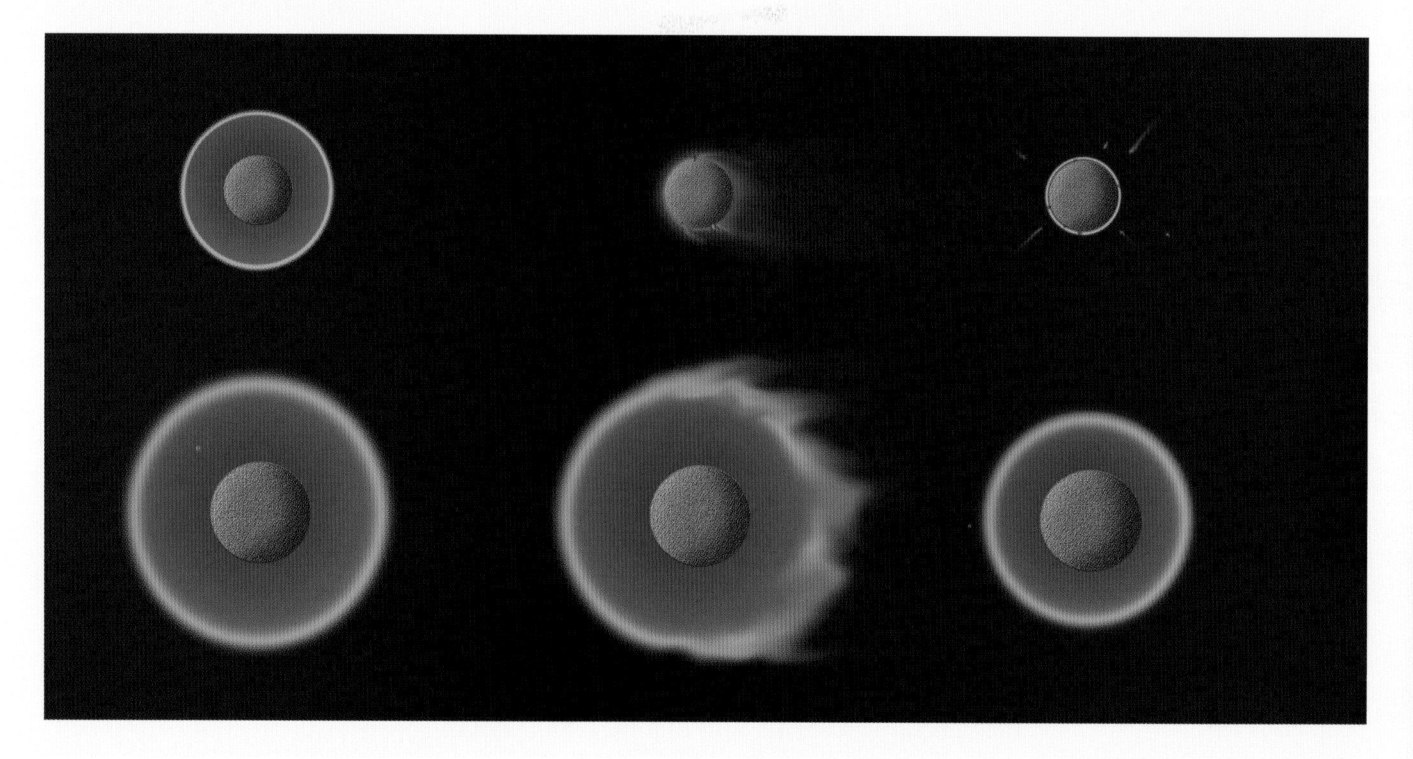

EVEN IF LIFE-BEARING WORLDS ARE ABUNDANT, ODDS ARE THAT THEIR INHABITANTS MAY BE LESS LIKE US AND MORE LIKE SLIME MOLD OR DANDELIONS.

Right: *Early in the solar system's formation, our Sun passed through an energetic phase called the T-tauri phase. Most stars go through a similar, violent era early in their development.*

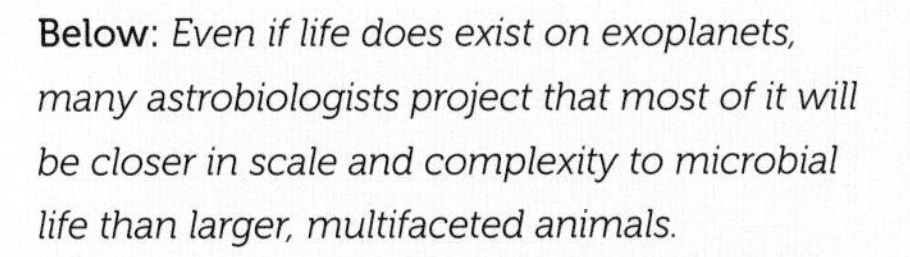

Below: *Even if life does exist on exoplanets, many astrobiologists project that most of it will be closer in scale and complexity to microbial life than larger, multifaceted animals.*

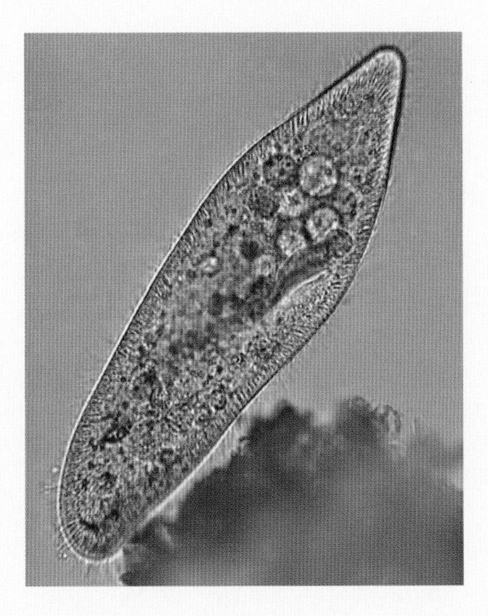

Sun was undergoing an active period called the T-tauri phase (see Chapter 3). During this epoch, the Earth lost 10 times the amount of atmosphere it has today. Volcanoes and some cometary material replaced that early atmosphere. Even today, during the occasional solar flare let loose by our Sun, the Earth loses 10kg (22lb) of oxygen every second. The Sun is stable enough that these flares don't last long, but the flares on a red dwarf are a different matter. They are frequent and deadly, and planets orbiting in the nearby habitable zone are vulnerable not only to their radiation, but to the effects of their solar winds, which erode the atmospheres of even something as large as a Super Earth. Even on an Earth 2.0 in a red dwarf system, it may be hard to make life and keep it burgeoning.

Even if life-bearing worlds are abundant in the universe, odds are that their inhabitants may be less like us and more like slime mold or dandelions. Finding any kind of biology beyond Earth will illuminate for us the very nature and heritage of life, showing us that: (1) either life out there is related to that on Earth, and perhaps brought here by wandering comets or meteors, or (2) life got its start separately at another site—an independent biogenesis—demonstrating that life must be widespread throughout the universe. Either proposition is exciting.

CARBON-BASED LIFE

On Earth, the foundation of all that lives is the carbon atom. Carbon works well in living processes for several reasons. Its chemical nature allows it to bond easily with elements like oxygen, nitrogen, sulfur, iron, hydrogen, and others—all important pieces in the puzzle of life. Because of its molecular structure, it is compact, which makes it easy for enzymes to use it in metabolic processes. Carbon is also abundant in the universe. Because of its penchant for easy bonding, carbon assembles into long chains of complex molecules. These chains—called organic compounds—are necessary for life functions in terrestrial life-forms, from single-celled cyanobacteria to blue whales.

Life on our own planet is based on a foundation of about 20 amino acids, chains of carbon molecules used in life's biotic operations. Over 500 varieties of these carbon chains exist in nature. The discovery of life based on other amino acids would mark the discovery of creatures that had their genesis via a profoundly different path from ours. The trick of finding alien biology, in addition to studying an exoplanet's environment, is to discover divergent chemical patterns from those we see on Earth.

Biologists are hard-pressed to find any chemical alternative that are as life-friendly as carbon. The most often cited element, both for complexity and flexibility in a biological system, is silicon. Silicon has some properties that are similar to carbon. Silicon, for example, organizes into chains of molecules that

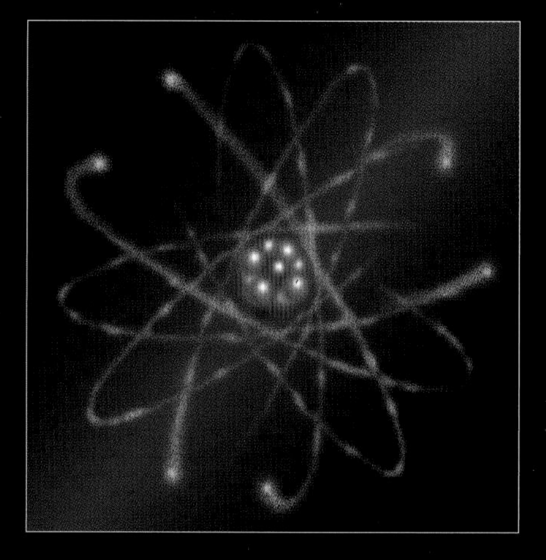

Above: *The versatile carbon atom forms long chains of molecules useful in life's processes. Its power comes from its ability to combine with many other atomic structures, creating an abundance of compounds like benzene, methane, and many proteins, to name a few. Carbon forms the spine of organic compounds necessary for life, interfacing with many different molecules.*

Left: *Organic compounds, including amino acids, are used by every form of life on Earth, from microbes to elephants. Even the simplest forms of life, lacking structures like nuclei and cell membranes, are carbon-based. At left,* Leucocoprinus birnbaumii *mushrooms are a popular addition to flowerpots. More mobile are African elephants, whose genome is based on the same foundation of organic compounds.*

Above: *Chains of carbon atoms are the key ingredient to all life on Earth. Seen here as indistinct clouds of subatomic particles, these complex atoms form amino acids and other organic compounds, the building blocks of life. Are alien life forms able to make use of other compounds?*

are large enough to carry out biological processes, just as carbon does. But silicon also has its limitations. Silicon's chemistry is not as adaptable as carbon's; it cannot bond with as many types of atoms. And unlike carbon, the way silicon forms bonds limits the kinds of shapes that its structures might form. Its molecules are large and bulky compared to carbon molecules, so they do not easily bond in collections common to organic chemistry. Still, it is found within Earth's biological processes.

Many carbon-based creatures incorporate silicon into skeletal or protective structures. Some biologists suggest that the arrangement of silicates in clays performed a crucial role in organizing carbon compounds during the formation of early life on Earth. Additionally, silicon compounds behave differently under conditions alien to those on Earth. At temperatures similar to those found on Saturn's moon Titan, for example, silicon polysilanes, related to sugars, are soluble in liquid nitrogen.

FINDING PARADISE

The exoplanets are painfully distant. Even the closest Earthlike one found, Proxima b, is so distant that the speedy New Horizons spacecraft would take over 50,000 years to arrive in its neighborhood. But we can tell some critical things about these far-off shores, and some of those things inform us of their habitability.

In their hunt for distant life-forms on exoplanets, scientists will search for biosignatures—imbalances in the environment. On our own planet, one of the most obvious biosignatures of life is the presence of oxygen. Oxygen is a reactive molecule: it wants to combine with others to create new elements. Because of this quality, it does not remain free in the atmosphere for long, but rather combines to create other forms (for example, the rust on iron is oxygen combining with materials in the metal). Oxygen must be replenished to remain at the levels that it does in our atmosphere. The photosynthesis of plants creates this important imbalance marking an active biome. But there are other ways of making oxygen. We have seen that early stars go through a T-tauri phase, with furious solar winds that can strip away an atmosphere. Often, that planetary atmosphere is recharged by volcanoes and comets. But the T-tauri process can

CHECKLIST FOR HABITABLE EXOPLANTS

Prerequisites	**Details**
Temperature	–15 to 122°C (5 to 252°F)
Water availability	Some days of precipitation or ~80% humidity
Light/geothermal energy sources	-0-
Radiation	50 Gy/hour (limits based on survival of *D. radiodurans* bacteria)
Nitrogen	Enough nitrogen for "fixed nitrogen" to be present
Oxygen	Over 0.01 atmospheres needed for complex life

Above: *Shopping list for life: a checklist of environmental items that might enable life to thrive includes water, temperature, light, and other factors.*

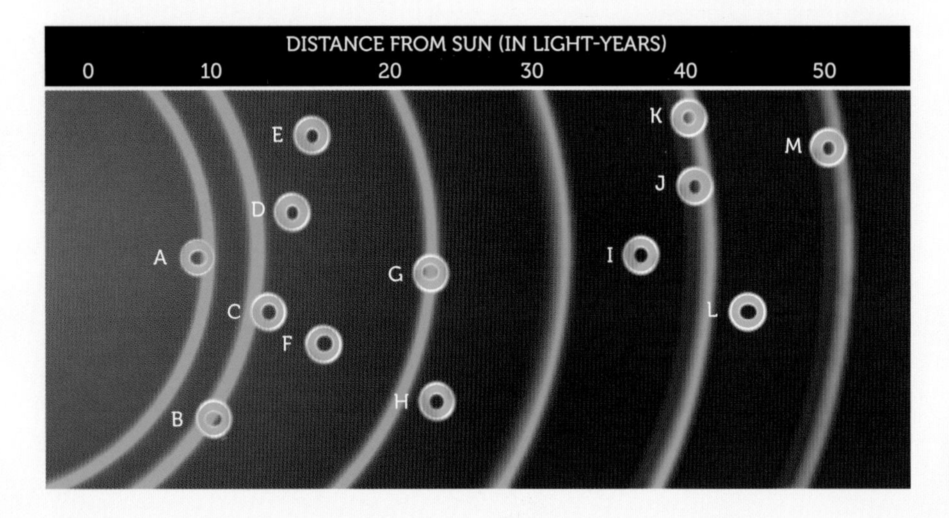

Left: *In our search for life in the universe, we will undoubtedly focus on our nearby neighborhood. Assorted close systems, including Alpha Centauri (which includes Proxima and the possibly Earth-like Proxima b), tally at least nine terran planets in the habitable zones of their stars. A=Proxima b (Alpha Centauri system); B=Tau Ceti e and f; C=Kapteyn b; D=Wolf 1061 c; E=Gliese 581 g; F=Gliese 682 c; G=Gliese 581 g; H=Gliese 667C c; I=HD 85512 b; J=TRAPPIST-1 d/e; K=Gliese 180 b and c; L=HD 40307 g; M=Gliese 163 c.*

= terrestrial exoplanets/ Super Earths in HZ

IN THEIR HUNT FOR DISTANT LIFE-FORMS ON EXOPLANETS, SCIENTISTS WILL SEARCH FOR BIOSIGNATURES—IMBALANCES IN THE ENVIRONMENT.

Above: *Plentiful water does not necessarily mean life, but it certainly is a prerequisite for life as we know it. Earthlike exoplanets and gas giants trailing terran exomoons will have the propensity for environments with water and—just perhaps—abundant life.*

also build up a large oxygen atmosphere that has nothing to do with life. Fierce radiation tears apart water molecules, separating them into oxygen and hydrogen in a process called dissociation. While the light hydrogen escapes into space, heavier oxygen remains behind. Finding oxygen on an exoplanet is not necessarily the smoking gun for life; oxygen needs to be replenished over time or it becomes chemically locked in a planet's rocks.

A recent study at Johns Hopkins University demonstrated that oxygen and organic compounds can both be generated in the atmosphere of an

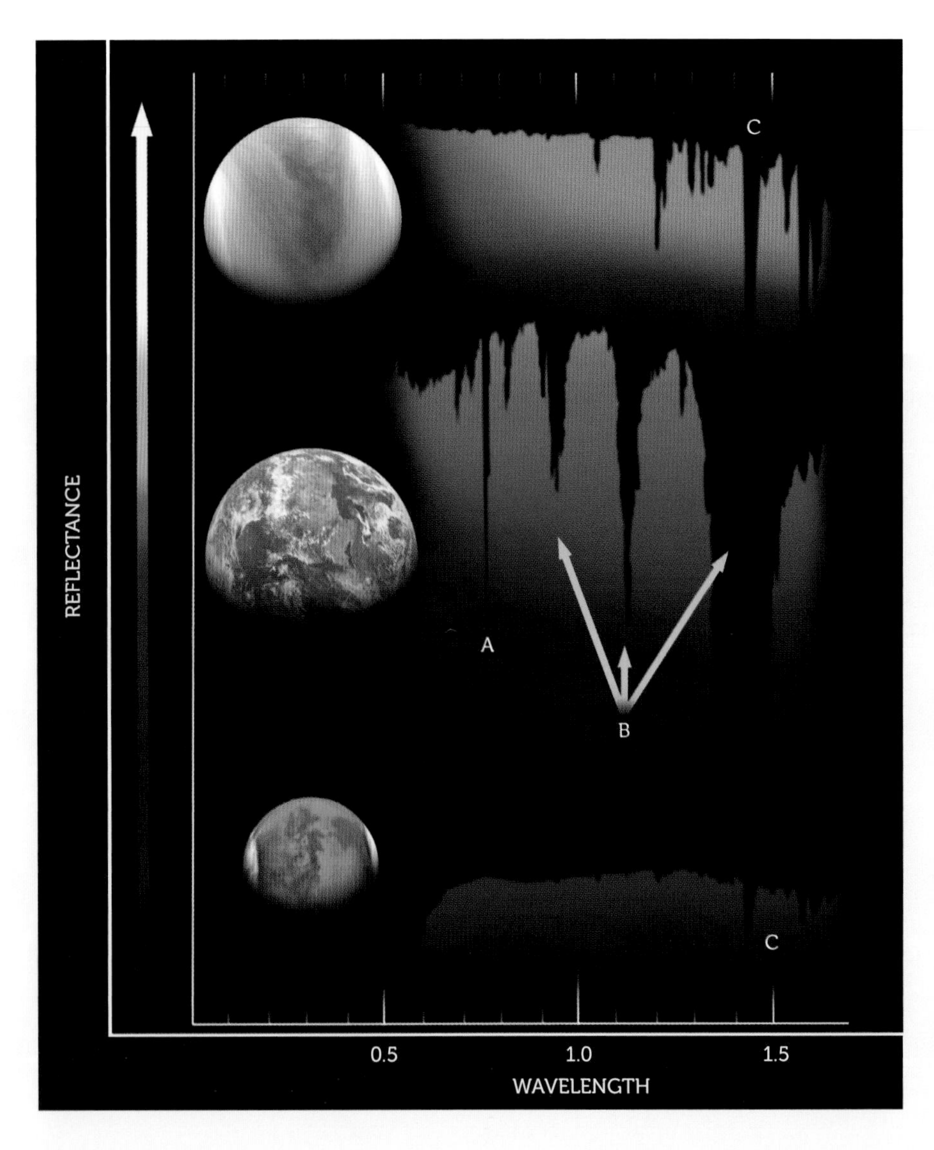

Left: *The spectra of light on three planets, compared. The light reflected by a planet's atmosphere and surface can tell us much about its makeup. A visitor to our solar system would see these three spectra. What could they tell? Note how complex the Earth's spectrum is compared to those of Venus (top) and Mars (bottom), both of which are dominated by carbon dioxide. Here, A indicates oxygen, B shows water, and C is the fingerprint of carbon dioxide.*

exoplanet in a lifeless environment. Researchers combined the gases thought to make up the air around Super Earths and sub-Neptunes. The study used carbon dioxide, methane, ammonia, and water vapor. When heated to temperatures between 25° and 370°C (80° and 700°F), the combination synthesized both complex organic materials and oxygen.

The test was a good cautionary tale for life searchers, but the presence of oxygen by itself would not confirm the existence of life even in Earth's environment, McKay says. "If you were to land on Earth and start measuring oxygen, you wouldn't see any variations. Even though there are organisms around you that are consuming it, their rate of consumption is tiny compared to the amount of oxygen in the air. The level would look rock steady." Any rover sniffing around for life signs would detect the same consistent levels of oxygen, whether it sampled the air in a house, in a jungle, or by a lake. But carbon dioxide is a different matter. Levels of carbon dioxide, for example, are highly variable.

ANOTHER CLUE TO ACTIVE EXOPLANETARY LIFE WOULD BE A LEVEL OF COMPLEXITY IN STRUCTURE OR CHEMISTRY . . .

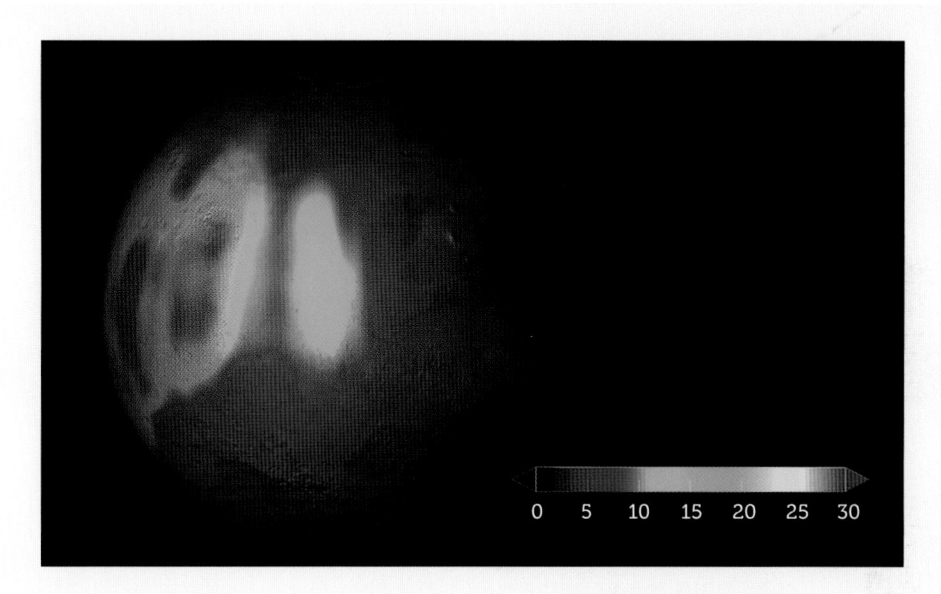

Right: *Methane levels appear to increase during the Martian summer in its northern hemisphere. This map shows global methane levels, with strong plumes over regions that have been affected by ancient ground ice or flowing water. The causes are unknown, but both geologists and astrobiologists are intrigued. Levels of this unstable gas usually indicate geologic activity (geothermal geysers or volcanic outgassing) or some kind of biogenic activity. In this diagram, which represents the Martian methane concentration in parts per billion, red indicates highest methane levels, and purple the lowest.*

"You measure it in a forest in the morning, and it would be very different than if you sat in that forest at night. It varies with seasons, location, and time." The reason for this changeability is the concentration of carbon dioxide in the atmosphere, which is very small compared to the total. The biologically ruled variation is large enough to affect the entirety of the atmosphere. If, for example, life on Titan consumes hydrogen the way that life on Earth cycles oxygen and carbon dioxide, it may be that it will also fluctuate. "It might vary near a shore or after a rain. We know the Huygens probe landed just after a rainstorm. That moistened ground should wake up—biologically—and start consuming hydrogen."

Methane may be a similar marker. It is a gas that tends to combine with others, so it cannot remain free in an atmosphere for long unless it is continually replenished. Usually, the presence of methane is a telltale sign of either volcanism or of metabolizing life. Methane has been found in the ices of Pluto and in the atmospheres of Titan and Mars.

Another clue to active exoplanetary life would be a level of complexity in structure or chemistry that differs from the background "noise" of geological organics. This kind of data is difficult to obtain remotely and will probably have to wait for advances in our technology or samples taken on site, which represents an even greater leap in our technology, as it involves sending something there. But visiting an exoplanet with one of our robot emissaries may not be so far in the future.

GETTING TO EARTH 2.0: THE BREAKTHROUGH STARSHOT MISSION

Interstellar travel was relegated to the writers of science fiction and the dreamers of the cinema. That is, until now. With the discovery of a potentially Earth-size planet in orbit around the nearest star, Proxima Centauri, engineers are joining the ranks of the dreamers, putting visions to paper and, in the very near future, perhaps to hardware. But sending a spacecraft there is a formidable task. Proxima b is 40 trillion km (25 trillion miles) from home.

Crossing that distance seems nearly impossible. Futurists speak of humans in suspended animation or generation ships where descendants of the engineers who launched the ships would finally make it to new worlds. Unpiloted craft are a different matter, however, and a team of engineers and scientists is bringing vision to reality in the form of a mission called Breakthrough Starshot. The Starshot concept is the brainchild of science billionaire physicist Yuri Milner. The mission scenario calls for the use of laser-propelled nanocraft, each the size of a postage stamp. Using solar sails driven by Earth-based lasers, each spacecraft-on-a-chip could make the journey in two decades, less than a generation. The technology-packed wafers, each at the center of their meter-wide gossamer laser-sail, would carry cameras, a tiny nuclear power source, a computer, and a communications system that would make use of the laser. Boosted by a network of lasers providing 100 gigawatts of energy, each craft could cross the orbit of Mars in two hours and Pluto in three days. In contrast, thanks to Jupiter's gravity slingshot, the Juno spacecraft reached an acceleration of 265,000km/h (165,000mph). At that speed, it would reach Proxima b in 17,160 years. Starshot is slated to make the journey in one-quarter of a century, traveling at about 20 percent the speed of light. At that speed, each craft would have brief moments to encounter its planetary target. But a stream of craft, with arrivals staggered over time, could chart changes in weather, geologic processes like volcanism, and even search for the biological "fingerprints" of living systems in the environment.

With its reduced flight time and advanced sensors, the flotilla of miniature spacecraft could fly by Earth's cousin in record time, and just might give us our first view of a planet hosting alien life.

Left: *The Planetary Society's successful solar sail project, LightSail 2, coasts over Baja California in this wide-angle view taken in July 2019. Similar sail technology would be used by Breakthrough Starshot's nanocraft.*

Right: *Light sails have been at the heart of many interstellar travel studies, including this version, a sail propelled from an Earth-based laser system. While this scenario envisions a gigantic human-tended ship, Breakthrough Starshot would utilize sails of about a meter across.*

Right: *The Breakthrough Starshot mission would use a battery of powerful lasers, beaming their energy from the Earth's surface, to accelerate its flotilla of micro-spacecraft to high speed. Engineers project that with long-duration pulses propelling the sails, speeds of 15–20 percent the speed of light can be reached.*

LOOKING FOR ISLANDS OF LIFE

Biosignatures provide researchers with the start of an itemized list for requirements for Earthlike life. As we have seen frequently in our survey of exoplanets, temperature is of primary importance. Temperature affects the availability of liquid water (too hot and only vapor is accessible; too cold and the water turns to solid ice). Temperature is a handy yardstick for exoplanet researchers stuck on Earth. It can sometimes be sensed directly—and remotely—with telescopes, and it can be estimated using climate and atmosphere models. We've charted many Hot Jupiters across many systems, but many of the terran exoplanets found so far endure high surface temperatures as well (Kepler-70 b and LHS 3844 b come to mind). Just what is the limit for organic life to survive?

Water is another item on the list. As a recent NASA Ames report put it, "Our understanding of life on exoplanets and exomoons must be based on what we know about life on Earth....Liquid water is the common ecological requirement." And there are other requirements.

TEMPERATURE

For high temperatures, larger animal life finds its limits in places the Sahara Desert or Death Valley, where temperatures can reach 57°C (135°F). But methanogens (methane-based microbes that live in water) can survive temperatures up to 122°C (252°F) where the water pressures are high. The intense pressure of the water prevents it from boiling. But as water is heated and maintained as a liquid under pressure, its characteristics change. It loses its ability to serve as a solvent and its interaction with dissolved biomolecules like lipids or nucleic acids is crippled. In a sense, water is robbed of its life-promoting qualities. Heated water in contact with rocks can also be efficient

Below: *A mosh pit of shrimp clamber over the stone columns of hydrothermal vents 2,300m (7,500ft) below the surface of the Caribbean Sea. Astrobiologists are investigating these high-pressure colonies, cut off completely from the Sun's energy, for insights into what forms alien life might take on other worlds like Europa, Enceladus, and icy ocean exoplanets.*

BUT JUST HOW MUCH WATER DOES LIFE NEED? THE "DRY LIMIT" OF LIFE IS OF INTEREST TO ASTROBIOLOGISTS.

Above: *Liquid water defines the habitable zone of a planetary system. Many exoplanets are water worlds, but their water exists in forms that may not be beneficial to life. On Earth, the water is skin-deep, pooling on the surface in a relatively thin film (in contrast to the hundreds-of-kilometer-deep oceans of some Super Earths). But it has a profound effect on our environment, balancing temperatures and enabling living systems to survive and prosper.*

in generating the chemical bonds used in biological reactions. In fact, this has been suggested as a source of life in the oceans of Saturn's moon Enceladus. Ecosystems on the Earth's seafloor, with their high temperatures and pressures, provide an example of possible life below the ocean of an exoplanet or exomoon.

WATER

We've seen that the habitable zone is defined by a region where liquid water can exist on the surface of a planet or moon. On worlds where the temperature is within that range, life may still be limited by the amount of liquid water that's available. Mars is a prime example. It lies on the outskirts of the Sun's habitable zone, but it is so small that its gravity is unable to hold on to a substantial atmosphere. Its thin air cannot keep temperatures much above the freezing point of water, and the pressure is so low that any liquid water making it to the surface instantly boils away as vapor. But just how much water does life need? The "dry limit" of life is of interest to astrobiologists. In dry environments, phototrophs (organisms that use light for energy) find refuge where water is held within and below rocks. Photosynthetic cyanobacteria and lichens thrive in dry deserts around the world and endolithic cyanobacteria live just below the surface of salty rocks in the heart of Chile's Atacama Desert. The only water they have available comes from water vapor. In the dry valleys of Antarctica, lichens form a green and black layer inside sandstone, getting their moisture from the melting of sporadic—and rare—snowfall. The conclusion: entire colonies of life can survive on even tiny amounts of moisture from rain, vapor, or rare melting snow.

ENERGY FOR LIFE

Living things can get energy from chemical reactions, geothermal (volcanic) heat, and the light of their star. We've seen many planets affected by tidal heating and their geological forces will provide much energy for biology, especially those planets in habitable zones. The light from

red dwarfs is different from that of K-type (orange) or G-type stars such as our Sun. But it is still usable by plants that carry out photosynthesis.

Aside from some exotic colonies huddling around hydrothermal vents, or microbes munching on minerals within the rocks of deep mines, photosynthesis forms the foundation for all life on Earth. Plants supply food, fuel, and oxygen to the creatures that inhabit our rich and diverse biome. Photosynthesis tints the spectrum of our planet, pumping oxygen into the air and shifting the surface colors because of land plants. The light spectrum of Earth has a green "bump" in it, but it also has an increase in red from the effects of chlorophyll in green plants.

Some life-forms are able to carry out photosynthesis under extreme conditions. Purple bacteria can perform photosynthesis using near infrared light,

Left: *Many terran exoplanets or moons lie outside of the main habitable zone, but they may still have deep waters beneath their surfaces. These environs can provide safe haven for biomes, where life may feed off of heat seeping out from the core of the planet or moon. 16 Cygni B b may be such a place. The planet is more massive than Jupiter, large enough to support a ring system, and certainly large enough to tow Earth-size moons that might host hidden seas inhabited by alien aquatic life. Studies show that any of its moons might have liquid water on the surface at some times of the year because its orbit is eccentric (egg-shaped).*

the kind of light that is stronger in red dwarf stars. Unlike Earth's plants, they don't produce oxygen. Lichens and other plants do not have a strong red edge in their spectrum like leafy Earth plants do, so they don't change the spectrum in the same way. Still, changes in the spectrum of exoplanets may be a good indicator of the presence of plants, or of microbes or even larger creatures that use photosynthesis. These changes in the spectrum will provide us with a good biosignature.

On a planet circling a red dwarf star, plants will look quite different, according to several studies. The dominant colors of plants under the light of an M star will be red or dark purple—that is, if they can survive. A NASA Goddard team estimates that microbes just getting started in a sea of a planet circling a red dwarf would need to live at least 9m (29ft 6in) underwater to escape the effects of flares. They would eventually develop structures and strategies for growing larger and surviving closer to the water's surface, perhaps even exiting the water.

If any photosynthesizing life inhabits planets near the giant blue F stars or the smaller K stars, they will likely show similar colors to those on Earth.

Above: *On planets orbiting G-type and K-type stars, photosynthesis will carry on in similar fashion to what we see on Earth, with green being the most probable dominant color (left). But plants living in the sunlight of an M star must incorporate a different part of the spectrum. The result might be leaves rich in colors near the red: orange, yellow, purple, and brown or black.*

Right: *The wild variety found throughout Earth's biomes may foreshadow similar diversity on other life-bearing worlds. Here, artist Joel Hagen envisions bioluminescent life in an ocean on a faraway, Earthlike exoplanet.*

STUDIES OF RED DWARF STARLIGHT AND PHOTOSYNTHESIS SHOW THAT PLANTS ON A RED DWARF EXOPLANET WILL LIKELY BLOSSOM IN BOUQUETS OF REDS, YELLOWS, AND ORANGES . . .

THE SEARCH FOR BIOSIGNATURES: AN ANTARCTIC METEORITE

The exoplanets are so far away that searching for biosignatures there is challenging. But we have had practice thanks to a grapefruit-size rock hidden within polar ices for 13,000 years. Our first modern hands-on test in the search for extraterrestrial life came with the investigation of the Allan Hills meteorite, ALH84001. Hardy field scientists discovered the space stone on Antarctic ice in the 1980s, but the rock wasn't really investigated until the 1990s. With great fanfare, a NASA team unveiled evidence of alien life within the Martian meteorite. Although their findings are still in dispute, their approach was well-reasoned and will likely be used in future searches. The geologists' work presaged remote studies that will have to be taken in our search for life among exoplanets. Among the biosignatures that scientists will search for are:

Chemistry and isotopes: Microbes often change their immediate surroundings by shifting the pH (acidity) or oxidation (think rust). If these shifts are found in microscopic isolated areas within the sample, they may in fact be due to biological activity. In the same way, microbes may leave a trail by their preference to certain isotopes (numbers of neutrons) in atoms. These minuscule "penchants" for one atom over another are sometimes documented in rock.

Organic material: Carbon compounds form the backbone of all known life-forms on Earth. Although they can be made by nonbiological processes, repeating patterns of organic material could be a red flag that life has been present.

Microscopic structures: The most difficult "fingerprint" of life to interpret is a physical structure. Extremely small fossils, called nanofossils, have been found in some terrestrial rock samples, but their forms (tubes, spheres, or chains of capsules) can be mimicked by non-biological processes. We have a long way to go in interpreting meteorites that have come from planets in our local neighborhood, let alone interpreting biosignatures among the exoplanets.

Above: *Microscopic images of the controversial ALH84001 meteorite found in the Allan Hills of Antarctica. The meteorite is a fragment of Martian crust. The image on the left shows what may be chemical concretions related to biological activity. At right are structures that may be remnants of microbes.*

WITH GREAT FANFARE, A NASA TEAM UNVEILED EVIDENCE OF ALIEN LIFE WITHIN THE MARTIAN METEORITE.

Below: *Left: A colony of thermophilic bacteria tints the moist ground orange at Mickey Hot Springs, Oregon. Thermophiles are responsible for many similar color variations at geothermal springs in Yellowstone, Iceland, New Zealand, and other sites. Center: These rod-shaped halobacteria thrive in salty environments. In fact, they cannot survive freshwater environments, but must have brine to carry out their life functions. Right: Tardigrades, or "water bears," can withstand harsh radiation. They are a subject of study by astrobiologists because so many Earthlike worlds orbit red dwarfs, where radiation levels can soar periodically.*

LIFE IN THE EXTREME

Exoplanets display a variety of harsh conditions, but life on Earth survives a wide spectrum of temperatures, pressures, radiation, salinity, and other environmental conditions. Microbes have been found in virtually every kind of environment that has water available. Creatures living in extreme environments are called extremophiles. They fall into several general categories: *mesophiles* grow in the moderate temperature range, from about 20°C to 45°C (68° to 113°F); *psychrophiles*, or cold-loving organisms, have been found at −40°C (−40°F) in films of water surrounding grains of sand. In arctic regions and Antarctica, psychrophiles thrive in under-ice streams and along melting ice edges. Astrobiologists study them as analogues to life on places like Europa and Enceladus.

Thermophiles, or heat-loving microbes, thrive at 50°C (122°F) or higher. They top out in environments with temperatures of about 70°C (158°F). A subset of thermophiles is called *hyperthermophiles*. These organisms enjoy 75°C (167°F) temperatures, the hottest environment of any microbe. Some hyperthermophiles at geothermal vents on the ocean floor tolerate temperatures up to 110°C (230°F).

Halophiles survive highly saline environments like the Great Salt Lake or the Dead Sea. They have a distinctive red tint from a light-detecting pigment that enables them to carry out photosynthesis. One of the biggest problems at red dwarf and other stars systems is radiation. Halobacteria seem to have beaten this problem, at least in part, by building a crust of salt that can shield them from some of the ultraviolet radiation. Salts are opaque to shortwave ultraviolet.

Though they may not like it, some creatures can tolerate high levels of radiation. The most famous of these is

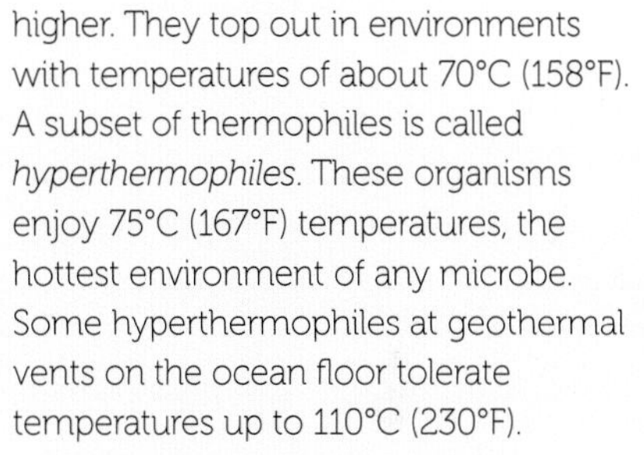

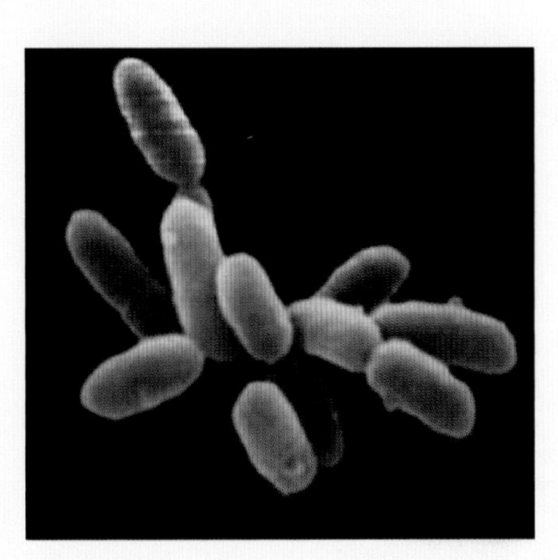

called the tardigrade, or "water bear." These tiny eight-legged creatures can survive levels of radiation 620 times the lethal dose for humans. A damage-suppressing (Dsup) film blankets their DNA, protecting them from the effects of X-rays, ultraviolet and other radiation. Dsup forms a low-density, "fluffy" shield, keeping protein-ravaging hydroxyl radicals from the proteins of the DNA. While they prefer sediment, moist soil, or moss, tardigrades live in diverse environments, and can withstand temperatures ranging from 149°C (300°F) down to –200°C (–328°F). Tardigrades, it seems, would do just fine at the limits of habitable zones around red dwarf suns.

Our survey of exoplanets is by no means complete. Many more worlds await discovery and our quest will undoubtedly reveal new kinds of planets and moons as well. But we have a good enough overview to begin considering what kinds of recognizable life may be out there, and where it may live.

Many of the planets we've found—both giant and terrestrial—circle red dwarf stars. Life there must endure the ravages of flares from their suns and it must be adapted to the bizarre effects of tidal locking. Studies of red dwarf starlight and photosynthesis show that plants on a red dwarf exoplanet will likely blossom in bouquets of reds, yellows, and oranges, many with purple to black leaves. Shores of heaving seas will witness great tsunamis as tides swell before the gravitational tug of nearby worlds.

Some gas giants may well have large moons, certainly qualifying as Earth 2.0s. Some life may encrust the landscapes of worlds in odd and eccentric orbits,

Above: *On planets in habitable zones, water will, in some places, be abundant. Life may be as well. Some gas giant worlds orbit in habitable zones of red dwarf suns. The larger planets likely trail larger moons, ones large enough to sustain oceans. Unlike their primary planets, the moons will not be tidally locked to their sun, but rather will witness true sunsets of their red dwarf sun.*

Below: *The surface of a planet in the habitable zone of a red dwarf star may be subject to the deadly radiation of periodic flares, but oceans can provide protection. Complex ecosystems may thrive beneath the surface of seas on exoplanets orbiting M stars.*

thriving as their planets near their suns, then clinging to life at the far end of the orbit, holding on until summer.

Farther out from their suns, creatures may ply the deep waters of ocean worlds. Many of these worlds will be Super Earths or terrans beyond the warm center of the habitable zone, but beneath their ice crusts may live riotous populations of swimming, floating, bobbing creatures, some living off light or radiation from their star, others converging on submarine volcanoes and the minerals they offer. More exotic materials may be involved in the operation of alien life. Some metals combine in ways similar to carbon. Titanium, tungsten, aluminum, magnesium, and iron can all form microscopic tubes, spheres, and geometric crystalline forms of the type found in diatoms. Metallic life might arise under conditions lethal to carbon-based forms. Even arsenic, deadly to carbon-based life, is incorporated into the biochemical functions of some organisms like algae and bacteria.

THE "RULE OF TWO" THEORY

In astrobiology, two extrapolates to infinity. Find just one more example of life beyond Earth (microbes count) and scientists will conclude it's all over the galaxy. But from there, it gets tricky. In 1996, a team of NASA biochemists and geologists announced the possible discovery of evidence of microbial life within a Martian meteorite known as Allan Hills ALH 84001. There is still much debate about whether the grapefruit-size meteor from Antarctica contains remnants—both chemical and physical—of Mars microbes. But if we find evidence of life on Mars, another question arises: Did that life originate on Mars? Or was it transported there on a similar meteor to Allan Hills, but going the other way? Cross-planet contamination is a distinct possibility. The Earth has meteorites that have been blasted from the surfaces of Mars, the Moon, the asteroid Vesta, and other sources. If we do find life on Mars or the moons of the outer planets, is it an offshoot of life here on Earth? Did life first arise on Earth, and then the Earth served as the genesis planet for all life in the solar system? Or, more controversially, what if life on Earth was planted here by meteors carrying life from Mars or other worlds? Some have even suggested that advanced races of aliens planted life here (as in Arthur C. Clarke's *2010*).

The other possibility is that sites like Mars, Europa, or Enceladus, if they have life, instigated a "second genesis," their own life in its own form. Perhaps, as some astrobiologists suggest, the universe is filled with life because it is filled with the building blocks of life: water, carbon, and other organic materials. These building blocks of life drift in the glowing nebulae, huddle in the rocks and ices of planets and moons, and even ride the winds of gaseous worlds.

The rule of two tells us that finding just one form of life outside of the Earth may mean that life is ubiquitous

Below: *In our own planetary system, astrobiologists believe the most likely sites for life are few in number. The water worlds offer deep oceans with rich resources. Closer to home, Mars is the most Earthlike terrestrial world in terms of environment. Still, it is a desolate place with temperatures usually well below freezing and air pressure equivalent to that on Earth at 30,750m (100,890ft). Left: Mars from the Curiosity rover, which has uncovered evidence of past standing water; Center: Europa's tortured crust hides a 100km (62-mile) deep ocean of salt water; Right: Saturn's moon Enceladus fires off water geysers 400km (250 miles) into its airless sky, making it a prime target for future life-seeking missions.*

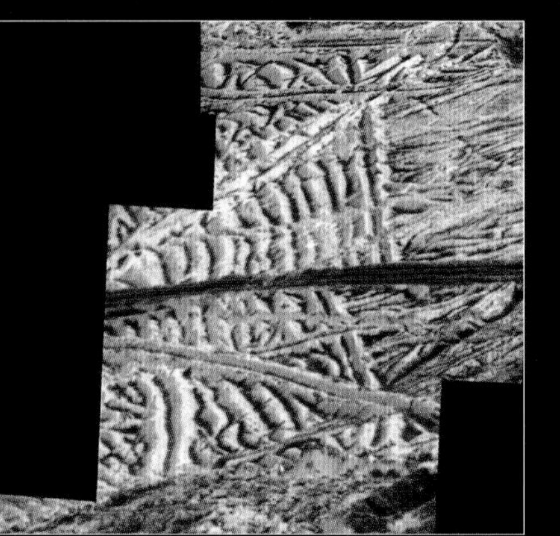

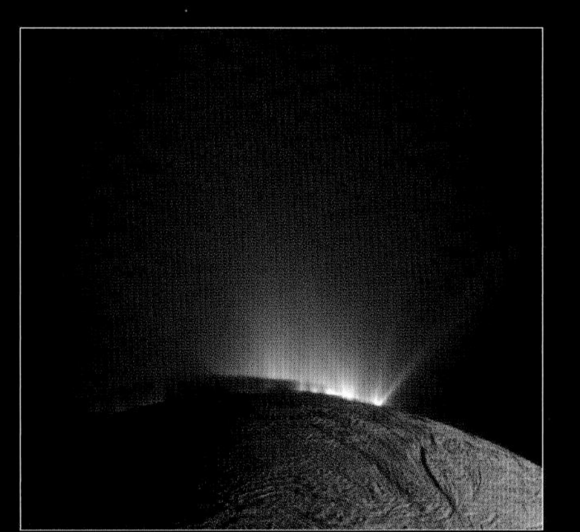

Right: *To possess Earthlike conditions on Proxima b, the planet must be far closer to its dim red dwarf star than Earth is to the Sun. The view from Earth (left) is compared to the view from Proxima b (right). In the Earth's sky, the Sun's diameter is half a degree. From Proxima b, far closer to its smaller star, the diameter of Proxima Centauri is three times that, or 1.5°. With this difference in the distance to their respective stars, temperatures are virtually identical for Proxima b and Earth.*

throughout the universe. We have found several worlds that might fit the moniker of "Earth 2.0." Even if the fraction of stars in the galaxy with potentially habitable planets is one out of 10, and if the stars in the Sun's neighborhood is typical of the rest of the galaxy, then finding just one of these Earthlike planets in a habitable zone shows us that the galaxy may have billions of Earthlike planets scattered across the Milky Way. But many researchers put the number of habitable planets in the trillions. What lives on them? Is something . . . or someone . . . looking back at us? We cannot know for sure. For now, we continue the search.

SEAS WITHOUT WATER

Could a world without water have life? Alien biology might survive in the places where water is under superpressure or where liquid ammonia serves as the river and sea of choice. We have an alien environment in our own solar system that may provide insights: the methane seas of Saturn's planet-size moon Titan (see "Breaking the stereotypes" feature on page 66).

Unlike Mars and Enceladus, where any Earthlike life will function in similar ways to terrestrial metabolism, finding life on Titan requires different tactics and tools. An initial probe would search for markers in the environment, like imbalances of gases or cycles of rising and falling gases that are tied with day and night. More advanced robots could land and rummage around looking for new structures or unusual, recurring, non-geological patterns. Orbiting spacecraft could detect biosignatures from above. Floating probes or submarines could sail the methane seas. One probe under construction, called Dragonfly, will monitor hydrogen levels while shoveling up surface samples, analyzing them with an advanced gas chromatograph/mass spectrometer. On water-bearing places like Mars and Enceladus, scientists are looking for water-based life, so they will search for molecules that work well in water like amino acids and lipids. On Titan, astrobiologists aren't yet sure what to look for. Researchers will be looking for anything that differs from the background environment.

WHILE WATER-BASED LIFE WILL LIKELY BE SIMILAR TO LIFE ON EARTH, ANY LIFE USING OTHER MATERIALS FOR METABOLISM (IN PLACES LIKE TITAN) WILL TELL US THAT MORE THAN ONE KIND OF LIFE EXISTS IN THE UNIVERSE.

ALIEN BIOLOGY MIGHT SURVIVE IN THE PLACES WHERE WATER IS UNDER SUPER-PRESSURE OR WHERE LIQUID AMMONIA SERVES AS THE RIVER OR SEA OF CHOICE.

Above: *Mars (left, seen from India's Mars Orbiter Mission) and ice ocean worlds like Jupiter's Europa and Ganymede (center), might have alien life, but that life will very probably be based on water, as life on Earth is. Should life be found on Titan (right, with light glinting off methane lakes), it will be truly alien. Titan's temperatures don't allow for liquid water, but liquid methane may take its place in biological processes there.*

The discovery of life on Titan would write a new chapter in the biological sciences. While water-based life will likely be similar to life on Earth, any life using other materials for metabolism will tell us that more than one kind of life exists in the universe, that there is a multitude of life and that life is incredibly diverse.

COULD A WORLD WITHOUT OXYGEN HAVE LIFE?

For 3 billion years, the Earth was similar to some Super Earth water worlds we have detected. Its atmosphere was mostly water vapor. The rest was carbon dioxide. The earliest life-forms—blue-green algae and other single-celled creatures—remained hidden within the oceans. These microorganisms used photosynthesis to build their bodies and carry out the business of biology. Using sunlight, they took in carbon by breaking down carbon dioxide from the air. The waste product was oxygen.

Like the cyanobacteria and blue-green algae, these tiny microbes and microscopic plants changed the atmosphere, bit by bit, of an entire planet. For billions of years, they flourished even while oxygen levels were undetectable. But eventually that changed. Over a thousand million years, they used up the carbon dioxide, replacing it with oxygen. The oxygen joined the nitrogen gas already in the skies. As the air changed, so did the life, adapting to the new gases and making use of them in efficient ways.

Oxygen is seen as a possible fingerprint—or biosignature—for active biology. Oxygen is "chemically reactive." It reacts with other things and changes

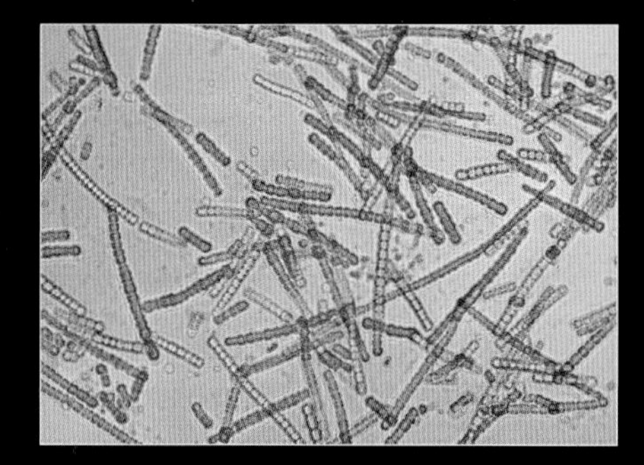

Above: *The tiniest creatures of the Earth, like the cyanobacteria and diatoms seen in this microscopic view of an algal mat from a marsh in Mexico, transformed the Earth's atmosphere from a carbon dioxide one to a life-sustaining canopy of nitrogen/oxygen air. Today, Earth's atmosphere is 20 percent oxygen, a gas used by the majority of life on the planet.*

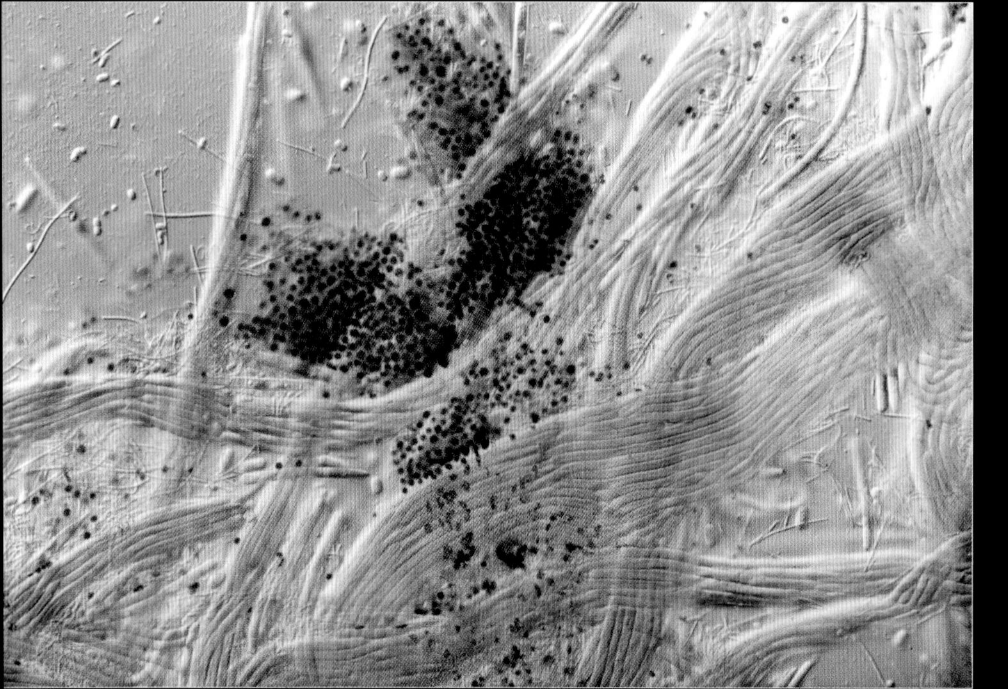

Left: *Microscopic view of blue-green algae from Guerrero Negro, Baja California Sur, Mexico. The oxygen-producing algae are the long greenish bundles, some of the oldest life-forms on Earth. Dark globes are sulfur bacteria.*

Above: *A world without oxygen: artist Walter Myers envisions a rich diverse biome on an exoplanet surrounded by an atmosphere devoid of oxygen. Astrobiologists suggest that such scenes may not only be possible, but plentiful among the exoplanets of our galaxy and beyond.*

its nature in doing so. Oxygen combines with iron, turning it to an orange powder in the process of rusting. It's what allows the fuel in a woodburning stove to flame, or causes the little explosions from gasoline in an engine's cylinders. It combines with hydrogen to make water. Oxygen reacts to so many things that it cannot remain free in the atmosphere for very long before it combines with something else. If there is oxygen in an exoplanet's atmosphere, something must be resupplying it.

Earth's early history proves that planets can host living biomes that do not depend on oxygen. Some processes can replenish oxygen, but the most powerful is the presence of living things.

FINAL DESTINATIONS

Life may take on many different and wondrous forms, some that we will recognize immediately as living organisms—perhaps something like those on Earth—and others alien enough that we may not realize they are alive at first. The sheer number of Earth-similar exoplanets tells us that across the galaxy may be a web of life spun from different silk, with its ebb and flow modeled in a completely foreign pattern. Our search for life in the galaxy has, up to now, been frustrated by great distances. Kepler has charted only one-ten-thousandth of 1 percent of our galaxy. TESS is continuing its full-sky survey, and advanced observatories like ESA's ARIEL and NASA's James Webb, will add to the ground we can cover—and the depth we can explore—in our pursuit of life in the galaxy.

Some astrobiologists suggest that the first places we should be looking are those most similar to our home planet because life there will likely have taken a similar path to the living things we understand here. But we have seen that Earth-mass planets in habitable zones may be rare. Still, they are out there, and with patience we will find them.

What might an Earth 2.0 be like? An evil twin? A "superhabitable" garden planet, bigger and better than our home planet? Or will it be a world on the edge of habitability, bleak and scarcely livable, a barely moistened Mars? Will we find life among the Earth 2.0s? If so, does this mean we keep "hands off"? Will that life have gained intelligence? Technology? Will they have religion, art, politics? Can we communicate with them, or will the crown of the alien evolutionary chain top out at something akin to dolphins and whales? Can we travel there, at least vicariously, with probes like Breakthrough Starshot?

Perhaps the greatest lesson we can learn in our search for Earth 2.0 is that our planet is a very special place in a critical location with a balance of many factors. NASA Goddard's Vladimir Airapetian points out that, "We have three planets in the habitable zone. Why is only one of the three habitable today? Is it something unique?"

Even if we find a world rich in life, the astronomical distances to even the nearest Earth candidates make travel impractical for the foreseeable future. But if our probes and telescopes find life out there, it will be a mirror of sorts, a looking glass to show us something about ourselves and the value of life on our own planet. Is life elsewhere like us? Does it represent a second genesis, or do we share a mysterious cosmic bond, a shared background lost in the darkness of deep time? If life has issued from a different source, what is our commonality? Our shared essence? The only certainty we have is that life—both on Earth and elsewhere—is precious, and the only life we currently know is on this planet. The vast wilderness of exoplanets beyond our solar system provides a contrast that inspires stewardship of our own world, our Earth 1.0.

Right: *An inhabitant of an exoplanet hibernates at the far end of its world's eccentric orbit, biding time until the sun nears and summer returns to warm its world. Some researchers suggest that life in the universe is a rarity, while others anticipate our future discoveries will uncover a universe teeming with such life, from creatures swimming below ice to herds of wondrous creatures only dreamed of.*

PERHAPS THE GREATEST LESSON WE CAN LEARN IN OUR SEARCH FOR EARTH 2.0 IS THAT OUR PLANET IS A VERY SPECIAL PLACE IN A CRITICAL LOCATION WITH A BALANCE OF MANY FACTORS.

FURTHER READING

The following books reflect resources about planets of other stars as well as books containing art of planets and exoplanets.

Bennett, Jeffrey and Seth Shostak. *Life in the Universe*. 4th edition. London: Pearson, 2017.

Carroll, Michael. *Ice Worlds: Their Tortured Landscapes and Biological Potential*. New York: Springer, 2019.

Carroll, Michael. *Earths of Distant Suns: How we Find Them, Communicate with Them, and Maybe even Travel There*. New York: Copernicus Books/Springer, 2017.

Carroll, Michael, and Rosaly Lopes, *et al. Alien Seas: Oceans in Space*. New York: Springer, 2013.

Flynn, Marilynn. *Space Art*. Self published: 2019.

Hardy, David A. *Visions of Space: Artists Journey through the Cosmos*. London: Paper Tiger, 1989.

Hartmann, William K. *Cycles of Fire: Stars, Galaxies and the Wonder of Deep Space*. New York: Workman, 1987.

Hobbs, Steven. *Imagining the Spheres: How We View Our Neighbouring Worlds*. Self published: 2018.

Miller, Ron. *Art of Space: The History of Space Art, from the Earliest Visions to the Graphics of the Modern Era*. New York: Zenith Press, 2014.

Perryman, Michael. *The Exoplanet Handbook*. 2nd edition. Cambridge: Cambridge University Press, 2018.

Seager, Sara, ed. *Exoplanets*. Tucson, AZ: University of Arizona Press, 2011.

Summers, Michael, and James Trefil. *Exoplanets: Diamond Worlds, Super Earths, Pulsar Planets, and the New Search for Life Beyond Our Solar System*. Washington, DC: Smithsonian Books, 2018.

Trefil, James and Michael Summers. *Imagined Life: A Speculative Scientific Journey among the Exoplanets in Search of Intelligent Aliens, Ice Creatures, and Supergravity Animals*. Washington, DC: Smithsonian Books, 2019.

EXOPLANET-RELATED WEBSITE RESOURCES

Planetary Habitability Laboratory at the University of Puerto Rico, Arecibo: http://phl.upr.edu/projects/habitable-exoplanets-catalog

NASA/Ames Center for Exoplanet Studies: https://www.nasa.gov/content/ames-center-for-exoplanet-studies-aces

NASA/Goddard's Exoplanets and Stellar Astrophysics Laboratory : https://science.gsfc.nasa.gov/astrophysics/exoplanets/

NASA Exoplanet Archive
https://exoplanetarchive.ipac.caltech.edu

NASA Exoplanet News
https://exoplanets.nasa.gov

European Space Agency Exoplanet Science
https://sci.esa.int/web/exoplanets

The Planetary Society's Exoplanet Research
https://www.planetary.org/explore/projects/exoplanets/

INDEX

Page numbers in *italics* refer to pictures and caption text.

16 Cygni B b *203*
47 Ursae Majoris b 32, *33*, 167
51 Pegasi b 32, *33*, 35, *105*
55 Cancri system 32, 95, 96–7
70 Virginis b 32, *33*

A-type stars 15, 140, 142
accretion disks 22, *23*
advanced (intelligent) life *181*, 185–6, 216
Ahlers, Johnathon 140, 142
air pressure 54
Airapetian, Vladimir 51, 188, 216
algae 214
alien life *see* extrasolar life
Allan Hills meteorite (ALH84001) 206
Alpha Centauri system 22, 87, 130
amino acids 190, *191*
ammonia 93, *95*
Antarctica 201, 206, 207
Arecibo radio telescope 19, 99
ARIEL (Atmospheric Remote-sensing Infrared Exoplanet Large-survey) 81
Aristotelian model 16
arsenic 209
asteroids 56
astrometry 30, 71, 79–80
Astronomical Unit (AU) 17
atmospheres
  air pressure 54
  climate models 107–9
  Earth 183, 189, *214*
  erosion/stripping *82*, 83, 107, 126, *146*, 189
  hot, rocky planets 106–7
  light spectra analysis 75, *77*
  Proxima b 137
  and tectonics 59–60
  TRAPPIST-1 system 122, 125
  ultraviolet light 142
  and volcanism 133
  water detection 162
Atmospheric Remote-sensing Infrared Exoplanet Large-survey (ARIEL) 81
AU *see* Astronomical Unit

B-type stars 15
bacteria 67, *182*, 201, 203–4, *207*, 209, 214
  see also microbes
Barnard's Star *127*, *128*
basalt 152
Beta Hydri *186*
Betelgeuse *13*, 14, *15*
binary-star systems 21, 22, 32, *92*, *144*, 161
biodiversity 54, *204*
biogenic zones 51
biological prerequisites 65–9
biosignatures 192, 200–6, 214–15
black holes 14, 22, *23*, 62, 63, 64
blue giants 13, *15*
blue-white dwarfs *15*
Borucki, Bill 71
Breakthrough Starshot Mission 196–9
brown dwarfs *142*
Bruno, Giordano 16–17

carbon 67, *185*
carbon-based life 182, 190–1
carbon dioxide 167, 194–5, 214
  CO2/rock cycle 167–8
Charon 112
CHEOPS (CHaracterising ExOPlanets Satellite) 80
civilizations *181*, 185–6, 216
climate models 107–9
clouds 96, *121*, *140*, 162
collapsing planets 147–52
comets 56, *67*
Copernican model 17
cores, planets 48, *51*, 54, *55*, 97
coronae *60*
coronagraphs 79
coronal mass ejections 136
  see also solar flares
Corot-7 b 95
CoRoT (Convection, Rotation et Transits planétaires) space telescope 73–4, 95
Cukier, Wolf 91
CVSO 30 system *43*
cyanobacteria *182*, 201

data challenges 170
Dimidium (51 Pegasi b) 32, *33*, 35, *105*
direct imaging technique 26, 43
disintegrating planets 147–52
dissociation process 193
Doppler effect 27, 130
Dorn, Caroline 95
double-star systems 21, 22, 32, *92*, *144*, 161
Draugr 19, *21*

Earth
  atmosphere 183, 189, *214*
  axial tilt 56
  climate regulation 167
  core 54, *55*
  habitable zone 163
  Kepler-20 system comparison *116*
  life-enabling conditions 54–62, 65
  location in Milky Way galaxy 65
  magnetosphere 54, *55*
  meteorites 206, 210
  and Moon 56–8, 112
  oceans 133
  orbit 54, 73, 163
  origin of life 66–7, 182, 183
  planetary companions 54–8
  plate tectonics 58–60, 167–8
  Proxima b comparison *211*
  recycling systems 58–61, *66*, 167–8
  volcanism *66*
  water/rock blend 54
  see also Super Earths
Earth Similarity Index (ESI) 66, 111, 171, 174
Earth-size planets 107, 115–16, 118, 188
Earthlike planets 82, 99, 100, 110–11, 123–5, 172–9, 211, 216
eclipses, solar *28*
Eddington, Arthur Stanley 28
"edge effect" 54
Enceladus 201, 210
Endl, Michael 130
energy, and life 201–4
ESI *see* Earth Similarity Index
Europa 182, 210
European Southern Observatory (ESO) 43

event horizon 22
exoplanets, number identified 22, 88
extrasolar life 181–217
  advanced life *181*, 185–6
  biosignatures 192, 200–6, 214–15
  Earth's local neighborhood *192*
  microbial life 183, 186, *189*
  on moons *203*
  non-water-based life 212–13
  optimism for 182
  oxygen biosignature 214–15
  rule of two theory 210–11
  water *193*, 209
extremophiles 128, 207
eyeball planets 112–17

F-type stars 15, *25*, 142
Fauchez, Thomas 118, 121, 122, 123, 126, 162
"flare stars" *see* red dwarfs (M-type stars)
Fomalhaut *26*, 76
Fomalhaut b *26*

G-type stars 12, *87*, *143*
Gaia mission 30, 79–80
Gamma Cephei system 22, 171
gamma-ray bursts (GRBs) 62–3
gas giants 38–40, 48, 87, 93, 152–4
  habitable zones 160–1
  HD 80606 b 164
  KELT-9 b 140–3
  moons 42, 168, 208
Gliese 436 b 147–8
Gliese 581 system 170–1
Gliese 667C system *35*, 91, *100*, *110*, *127*
Gliese 682 c *192*
Gliese 1214 b 161–2, *163*
Gliese 3470 b 146–7
Goldilocks zones *see* habitable zones
Grand Tack theory 38
gravitational microlensing 28–9, 81
gravity-darkened stars 142, *143*
GRBs *see* gamma-ray bursts
greenhouse gases 167

HabEx (Habitable Exoplanet Observatory) 81
Habitable Exoplanets Catalog 99
habitable zones 24–5, 51, 54, *65*, 100, *103*, 105, 115
  gas giants 160–1
  gravity-darkened stars 142, *143*
  migration of 163, 167
  moons 160–1, *203*
  red dwarfs *209*
  Sun 163, 168
  TRAPPIST-1 system *127*
halophiles 207
HAT-P-7 b (Kepler-2 b) 84
HAT-P-12 b *53*, 83–7
HATNet project 84
HD 40307 g *192*
HD 80606 b 164
HD 85512 b *192*
HD 188753 A b *38–9*, 170
HD 189733 b 156, *157*, *159*
HD 222582 b *165*
helium 13, 14, 48
Hot Jupiters *11*, *41*, 42, 82–3
  51 Pegasi b 32, *33*, 35, *105*
  HAT-P-12 b 83–7
  HD 189733 b 156, *157*, *159*
  Jupiter 38
  moons 156
Hot Neptunes *37*, 147–8
Hoth (OGLE-2005-BLG-390L b) *168*
HR8799 system *26*
HR8832 system 91–3
Hubble Space Telescope *73*, 74–7, 79, 156
human curiosity/exploration 11
Huygens, Christiaan 31
hydrocarbons 67, *68*
hydrogen 12, 13, 14, 48, *147*, *185*, 193
hydrothermal vents *200*

infrared radiation 105
infrared telescopy 76, 81
intelligent (advanced) life *181*, 185–6, 216
interferometry 26
Io 96, *97*

James Webb Space Telescope (JWST) 78–9
Juno spacecraft 38, *41*, 196
Jupiter *30*, *33*, 38–40, *41*, 48, *57*, *156*
  asteroid/comet interaction 56, *67*
  clouds 96
  daily rotation 87
JWST *see* James Webb Space Telescope

K2-18 b *74*, 162
K2-22 b *149*, 150–1
K-type stars 14, *22*, 91
Kappa Ceti *24*
KELT-9 system 140–3
Kepler-2 b 84
Kepler-10 b 85
Kepler-13 system 140
Kepler-16 AB b *179*
Kepler-20 system 116, 117
Kepler-35 b *90*, 91
Kepler-62 system 174, *177*
Kepler-64 system *87*
Kepler-70 system 152–5
Kepler-186 system 173–4, *174–5*
Kepler-442 b 174–9
Kepler-452 b *105*, 172–3
Kepler Space Telescope 29, *30*, 37, 53, *69*, 70–3, 77, 82, 88, 116, 216
Kopparapu, Ravi 24, 105
Kuiper Belt 131–2

Lalande 24
lasers *198*
LHS 3844 b *150*, 151–2
life
  origin of 182, 183, 189
  and radiation 207–8, *209*
life-enabling conditions 54–63, 65–9
  checklist *192*
  climate 107–9
  galaxy type 62–3
  planet size 115
  star type 61–2, *144*
  see also habitable zones
light
  blue/red shift 27
  spectra analysis 75, *77*, *194*, 204
LightSail 2 196
Lopez, Eric 38, 48–51, 60, 146, 147
LTT 1445 A b 158
LTT 9779 b *36–7*, 37, 38
Lynx X-ray Observatory 81

M-type stars *see* red dwarfs
McKay, Chris 182, 194, 195
magma oceans 95
magnetic fields *55*, *103*, 136
magnetospheres 54, *55*, *103*, 136
main sequence stars 13, 15
Mandell, Avi 51, 77, 82, 87, 96, 150–1, 164
Mars
  axial tilt 56
  life on 201, 210, *213*
  methane levels *195*
  tectonics 60–1
  volcanism 60, *61*
mass, planets 27, 30, *51*, 80, 82, 130–1
Mercury
  lava plains *151*
  orbit 135
  temperature 152
metallic life 209
meteorites 206, 210
meteors 182
methane 67, 68, *95*, *147*, 195, 212
Methuselah (PSR B1620-26 b) *144*, 146
micro-spacecraft 196–9
microbes 183, 186, *189*, 200, 204, 206, 207, 214
  see also bacteria
microfossils 183
microlensing technique 28–9, 81
Milky Way galaxy 63, 64–6, 88
Miller–Urey experiments 182
minerals 60, 65
Moon 56–8, 112, *151*
moons 42–7, 66, 128–9
  extrasolar life *203*
  gas giants 168, 208
  habitable zones 160–1
  Hot Jupiters 156
mountain chains 59
Mu Cephei *15*
multiple-star systems 87–91, 144

naming conventions 34–5
nanocraft 196–9
nanofossils 206
nebulae 152–5, *185*
Neptune
  core 48, 51
  and Kuiper Belt 131–2
  moons 45–7
neutron stars 14, 15, 19, 22, *63*
NGC 6217 (galaxy) 62
NGC 6744 (galaxy) *88*
NGC 6791 (galaxy) *30*
Nice Theory 38
nitrogen *47*, 185

O-type stars 15
oceans *51*, 125–6, 128, 133
OGLE-2005-BLG-390L b (Hoth) *168*
orange dwarfs 14
orbits 82, *115*, 135, 137, 140
  Earth 54, 73, 163
  shape of 164–5
organic compounds 190, 206
oxygen 67, 133, *182*, *185*, 192–4, 214–15

parallax 79–80
Phobetor *18*
photometry *see* transit technique
photosynthesis 67, 203–4, 208
Pi Mensae system *139*, 158–61
planet-hunting 18–31
  astrometry 30, 71, 79–80
  data challenges 170
  direct imaging technique 26, 43
  gravitational microlensing 28–9, 81
  radial velocity method 27, 31
  transit technique 29, 31, 70–3, 82
  transit timing variation technique 29–30
  see also habitable zones
Planetary Habitability Laboratory *109*
planetary relocation 131–2
planetary systems, evolution of 42, 51
planets
  cores *51*, 54, *55*, 97
  density 31
  eyeball planets 112–17
  formation of *12*, 97
  mass 27, 30, *51*, 80, 82, 130–1
  orbits 54, 73, 82, *115*, 135, 137, 140, 163, 164–5
  protoplanetary disks *12*
  retrograde motion 112
  see also individual planets
plants 203, 204
plate tectonics 58–61, 167–8
ploonets 42–7
Pluto 112, 132
precession 56, 58
prominences, solar *134*
proplyds 76
Proxima b 22, *99*, 130–7, 192, 196
  Earth comparison *211*
  habitability of 134–6
  mass 130–1
  migration of 131, 132
  orbit 137
  radius 133
Proxima Centauri *15*, 87, 134, 136
PSR B1257+12 19–21
PSR B1620-26 system 21–2, *144*, 146
psychrophiles 207
pulsars 14, 19–22, *144*, 154

Quintana, Elisa 132, 173

radial velocity technique 27, 31, 105
radiation, and life 207–8, *209*
recycling systems 58–61, *66*, 167–8
red dwarfs (M-type stars) *9*, 12, 13, 14, *15*, *22*, *24*, *25*, *34*, 51, 99–138
  brightness variability 73
  extrasolar life *204*
  eyeball planets 112–17
  habitable zone *103*, 105, *209*
  lifespan 107
  magnetosphere *103*
  planetary fates *188*
  solar flares 61–2, *72*, *73*, *103*, 134
  temperature 105
  tidal locking 112–27
red giants *13*, 14, *15*, 22
retrograde motion, planets 112
ruby and sapphire vapor 84
rule of two theory 210–11

Saturn 38, 48, 87
seasons 142
Shoemaker–Levy 9 (comet) *67*
silicon 190–1
singularity 22

Sirius *15*
sodium 96
solar eclipses *28*
solar flares 61–2, *72*, *73*, *103*, 134, 189
  see also coronal mass ejections
Solar System 8, *25*, 38, 54–8
  early development of 188–9
  light spectra analysis *194*
  moons 128
  orbit of 65
solar winds *82*, 83, 189
Spitzer Space Telescope 74, 76–7, 151, 156
stars
  A-type 15, 140, 142
  B-type 15
  bending of light 28
  binary-star systems 21, 22, 32, *92*, *144*, 161
  blue giants 13, *15*
  blue/red shift of light 27
  blue-white dwarfs *15*
  F-type 15, *25*, 142
  formation of 12
  G-type 12, *87*, *143*
  gravity-darkened stars 142, 143
  K-type 14, *22*, 91
  life-enabling conditions 61–2, *144*
  lifespan 13
  main sequence stars 13, 15
  naming convention 35
  neutron stars 14, 15, 19, 22, *63*
  O-type 15
  pulsars 14, 19–22, *144*, 154
  red giants *13*, 14, *15*, 22
  T-tauri phase 122, 134, *189*, 192–3
  types of 12–15
  white dwarfs 14, 154
  see also individual stars; red dwarfs (M-type stars)
steam worlds 51
stromatolites *182*
subterran planets *103*
sulfur *48*, 96, *97*
Sun 12, 14, 15, *30*, *33*, *61*, 62
  eclipse of *28*
  habitable zone 163, 168
  solar flares *134*, 189
Super Earths 48–51, 60, 74, 91, 92–3, 95, *99*, 110, *137*, 158–69, 188
  Earthlike planets 172–9
  formation of 97
  orbits *115*
  water detection 162
supernovae 14, 15, 64
superrotation *108*, 109
superterrans 111, *132*

T-tauri phase, stars 122, 134, *189*, 192–3
tardigrades *207*, 208
tectonics 58–61, 167–8
Teegarden system 100, 102–7
telescopes/observatories
  Arecibo radio telescope 19, 99
  glare removal 26
  Hubble Space Telescope *73*, 74–7, 79, 156
  James Webb Space Telescope 78–9
  Kepler Space Telescope 29, *30*, *37*, 53, *69*, 70–3, *77*, 82, 88, 116, 216
  Lynx X-ray Observatory 81
  Spitzer Space Telescope 74, 76–7, 151, 156
  TESS (Transiting Exoplanet Survey Satellite) 37, 78, 91
  Very Large Telescope 43
  Wide Field Infrared Survey Telescope 81
temperature
  CO2 levels 167
  extrasolar life 200–1
  Mercury 152
  thermal mapping 156
terminator zone *77*, 112, *113*, *115*
terran planets 95, *100*, 111, 116, *192*, *203*
TESS (Transiting Exoplanet Survey Satellite) 37, 78, 91
theories
  Grand Tack theory 38
  Nice Theory 38
  rule of two theory 210–11
thermal mapping 156
thermophiles 207
tidal heating 96, 160
tidal locking 107, 112–27, 135, 171
tides 58
Titan 67–8, *108*, 212–13
TOI 1338 b 91
transit technique 29, 31, 70–3, 82
transit timing variation (TTV) 29–30
transits, planetary 31, *72*, *75*, *77*, *148*
TRAPPIST-1 system 34–5, 116, 118–27, *192*
TrES-2 b *84*, 85
triple-star systems *38–9*
Triton *46–7*, 47
TTV *see* transit timing variation

ultraviolet (UV) light 142, *154*
Upsilon Andromedae system *160*, 161
Uranus 48, 51
UV (ultraviolet) light 142, *154*

van Maanen, Adriaan 18
Venus
  atmosphere 48
  core *55*
  daily rotation 112
  Kepler-20 e and f comparison *116*
  superrotation *108*
  tectonics *60*, 61
  temperature 48
Very Large Telescope (VLT), ESO 43
volcanism 60, *61*, *66*, 85, 96, *121*, *132*, 133, *150*, *151*, 152, 160

WASP-49 b 96–7
WASP-121 b *11*, 82–3
water 25, 51, 54, 83, 105, *121*, 126
  detection of 162
  extrasolar life *193*, 200–1, 209
  non-water-based life 212–13
water cycle 58, *104*
water worlds 48–51, 66, 111, 115, *133*, 174, 209
Wells, H. G. 185
white dwarfs 14, 154
Wide Field Infrared Survey Telescope (WFIRST) 81
wind circulation *47*, *108*, 109, 156
Wolszczan, Alex 18–19

X-rays 81, 107, 134, *159*

Youngblood, Allison 100, 107, 136

Zeta Leonis *24*

ACKNOWLEDGMENTS

First and foremost, my thanks goes to our tireless resource person, Goddard Space Flight Center's Elisa Quintana, who gave many hours of her personal time to see to the accuracy of this book. Any inaccuracies are due to the author's overactive creativity. And while Elisa was the maestro behind content, her counterparts in the book process were the talented Elephant Books team, starring Laura Ward, the tireless Chris McNab and Will Steeds, along with brilliant designer Nigel Partridge. My thanks to them for enabling the process to unfold as it should, and to Chris for making it so entertaining. Dr. Caroline Dorn at the University of Zurich advised me on a world of vaporizing rock and sapphire-clouded skies, not an easy thing to envision. Professor Abel Mendez, Director of the Planetary Habitability Laboratory in Puerto Rico, contributed images and advice. My thanks to Allison Youngblood, who found time to talk about M stars even in the midst of a move from one coast to the west. Many people at various NASA facilities helped in this project, including Chris McKay, Eric Lopez, Avi Mandell, Johnathon Ahlers, Ravi Kopparapu, Thomás Fauchez, Michael Lentz and Padi Boyd.

I extend my heartfelt appreciation to talented photographer/artist Rosemary Clark, who generously lent her beautiful cloud images for some of my digital works in this book. A special shoutout to my photographer/writer friend John Vester as well, for spectacular craggy alpine landscapes. And finally, to the many artists who generously contributed to the collection of exoplanet visions represented in this book (with special commendation to Ron Miller for help that went far beyond the art, and to Garry Harwood and Mark Garlick for spiffy caption material that made me sound as if I knew what I was talking about).

For this project in particular, my gratitude goes to Andrew and Xandra Carroll for helping out with visuals and moral support. And as always, to my bride and best friend, Caroline, for being there when I needed her.

A WORD ABOUT THE PAINTINGS

The science of exoplanets presents many moving targets for the artists attempting to portray distant worlds. Usually, researchers can tell us a range of possibilities for a planet, but the artist must settle on one. For example, Proxima b, the closest Earthlike planet to ours, orbits the red dwarf M star Proxima Centauri. Estimates for its size vary, as do estimates of its distance from its sun. These and other factors give us a range of possible conditions on the planet. Is it a huge Super Earth? A water world? A planet with Venusian clouds and a toxic, dense atmosphere? Astronomical artists are called upon to show all of these planetary alternatives, and they do so with gusto. Within these pages we see three general types of paintings. Traditional works are typically done in oils, acrylics, or other traditional media on illustration board or—more rarely—canvas. Digital work uses the same rules of composition, light, color mixing, and so forth, but paints with pixels using programs like Vue, Terragen, Photoshop, and Blender. Tradigital pieces combine the two with a traditional base imported into the computer and then enhanced by various programs used by digital artists. Hence, this book presents the best of those three worlds.

PICTURE CREDITS

The publisher would like to thank the following people and organizations for providing the images in this book. All artworks are © Michael Carroll with the following exceptions. Page number and position are indicated as follows: T = top, L = left, R = right, C = center, CL = center left, CR = center right, B = bottom, BL = bottom left, BR = bottom right, etc.

Don Davis: 58; Dan Durda, FIAAA: 168: T; 178: T; 193; 208; © 2020 Dirk Terrell: 87; © Pamela Lee: 217; Joel Hagen: 204–05; 209; Dr. Mark A. Garlick / space-art.co.uk: 14–15; 100–01; 138; 145; © William K Hartmann: 12: L; 45; © Lucy West: 72: T; © Raymond Cassel: Half-title page; Marilynn Flynn / Tharsis Artworks: 46–47; 97: L; © David A. Hardy / www.astroart.org: 85; 128; 171: L; Gemini Observatory/AURA/Lynette Cook: 155; © Lynette Cook: 18; 90; 165; Jett McIntyre Furr: 83; Steven Hobbs: 86; Ron Miller: Cover; 2; 38–39; 40; 95; 104–05; 106; 129; 152–53; 158; 160–61; 166–67; 168–69; 186–87; 202–03; Garry L. Harwood: 154: L; © Walter B. Myers: 215; Fahad Suleria, a.k.a. Tyrogthekreeper: 22; ESA/Hubble, M. Kornmesser: 74; NASA/Ames/W. Stenzel: 105: R; ESA/L. Calcada: 110; NASA/JPL-Caltech: 127: B; 150 Bill Gerish: 133; 201; NASA/CXC/SAO/K. Poppenhaeger et al: 159 (X-ray); NASA/CXC/M. Weiss: 159 (main illustration); Elisa Quintana/NASA: 6–7; Jastrow: 16; Blackcat/ Sergio D'Afflitto/CC-BY-SA-4.0: 17: BL; Public domain: 17: TR; LOC/PD-US: 28: TL; C. Davidson/PD: 28: TC; ESO/ Landessternwarte Heidelberg-Königstuhl/F. W. Dyson, A. S. Eddington, & C. Davidson/CC-BY-4.0: 28: TR; NASA/ Troy Cryder Feb. 16, 2009: 69; NASA Ames/Kepler Mission/Jason Rowe/Chris Burke/Wendy Stenzel: 70; NASA: 73; PHL @ UPR Arecibo (phl.upr.edu) Sep 4, 2019: 109; NASA/JSC: 112; Zdenek Bardon: 130 (visible light background); NASA/CXC/ Univ. of Colorado/T. Ayres et al.: 130 (X-ray inset top); STScI/ESA: 130 (lower inset); NASA/Johns Hopkins University Applied Physics Laboratory/Carnegie Institution of Washington: 135; 151: BL, BR; NASA/Ames: 182; Henrique Alvim Corrêa (PD): 185; Deuterstome CC BY-SA 3.0: 189: BL; lorek85 GNU: 190: BL; Xandra Carroll: 190: BC; Courtesy Chris German, WHOI/NSF, NASA/ROV Jason © 2012 WHOI: 200; Bill Gerrish: 201; NASA: 206: L, R; 207: C; 214: L; Amateria1121/CC BY-SA 3.0: 207: L; Schokraie E, Warnken U, Hotz-Wagenblatt A, Grohme MA, Hengherr S, et al. (2012)/CC BY 2.5: 207: R; SRO/ISSDC/Emily Lakdawalla: 212; NASA/ JPL/Caltech: 213: L; NASA/JPL-Caltech/ University of Arizona/University of Idaho: 213: R; Willem van Aken, CSIRO/CC BY 3.0: 214: TR; NASA/ JPL: 108: BL; NASA/ JPL/Space Science Institute: 108: BR. KU Leuven IvS/CmPA/Ludmila Carone: 108: T.

ARTIST WEBSITES

Andrew Carroll: https://www.andrews-world-of-art.com
Michael Carroll: www.stock-space-images.com
Ray Cassel: https://www.artstation.com/raycassel/challenges
Lynette Cook: https://extrasolar.spaceart.org
Don Davis: www.donaldedavis.com
Dan Durda: www.3dimpact.com
Marilynn Flynn: www.tharsisartworks.com
Jett McIntyre Furr: https://aerojettfurr.myportfolio.com/projects
Mark Garlick: www.markgarlick.com
Joel Hagen
David A. Hardy: www.astroart.org
William K. Hartmann: https://www.psi.edu/about/staff/hartmann/pic-cat/399-300.html
Garry Harwood: www.garryharwood.co.uk
Steven Hobbs: www.stevenhobbsphoto.com.au/index.html
Pamela Lee: https://pamelaleespaceart.com
Ron Miller: www.black-cat-studios.com
https://spaceart.photoshelter.com/index
Walt Myers: www.arcadiastreet.com
Dirk Terrell: www.boulder.swri.edu/~terrell/dtart.htm
Lucy West: www.lucyweststudios.com